ENGINEERING
AND
ENVIRONMENTAL
ETHICS

ENGINEERING AND ENVIRONMENTAL ETHICS

A Case Study Approach

Edited by

JOHN R. WILCOX
LOUIS THEODORE

JOHN WILEY & SONS, INC.
New York · Chichester · Weinheim · Brisbane · Singapore · Toronto

This publication is designed to provide accurate and authoritative information in regard to the subject matter covered. It is sold with the understanding that the publisher is not engaged in rendering professional services. If professional advice or other expert assistance is required, the services of a competent professional person should be sought

Library of Congress Cataloging-in-Publication Data:

Engineering and Environmental ethics : a case study approach / edited
   by John R. Wilcox, Louis Theodore.
      p.   P:020.
   Includes bibliographical references and index.
   ISBN 0-471-29236-2
      1. Environmenal engineering—Moral and ethical aspects.
   2. Engineering—Moral and ethical aspects.   I. Wilcox, John R. (John Richard,
   1939– . II. Theordore, Louis.
   TD153.E54   1998
   179'.1—dc21   97-38063

Printed in the United States of America

10 9 8 7 6 5 4 3 2 1

To
SUZANNE
KENNEY
LILLIAN
CHRIS
of whom I am so proud
(JW)

and to
TONY
JOHN
DON
whose successes have served as an inspiration and whose
friendship and support have touched my heart
(LT)

Contents

Preface

Over the past several years, the editors
have used case studies to discuss ethical issues in the
engineering and scientific professions. Having done this
separately in their respective course offerings and jointly
in a chemical engineering seminar, both believe that the
case method is one of the best ways to engage students
and professionals in the discussion of moral challenges
facing the engineering and scientific communities. The
editors also feel that individuals are more committed to
that discussion if they write up cases themselves.

This case study anthology is the result of the editors'
beliefs, and the enthusiasm and imagination of the case
authors, all of whom were students in the School of
Engineering at Manhattan College. Approximately half
the cases address ethical issues in environmental engineering. They are categorized under five headings: air,

solid waste, and water pollution; domestic applications; and health, safety, and accident prevention. The remainder of the cases address issues in four engineering fields: chemical, civil, electrical, and mechanical engineering.

Each case provided in the body of the text is followed by questions for discussion. These questions are by no means definitive. While they will help individuals focus on the case, the issues raised will make the most sense if they lead to a wide-ranging discussion among all readers. Analysis of ethics cases comes alive in group work. Answering the questions individually is a helpful first step, but one's understanding of ethical problems and dilemmas improves dramatically in group discussion.

This book is a valuable resource for both practitioners and academics. The United States Federal Sentencing Commission provides a reduction in sentencing for organizations that have established ethics training programs in order to reduce wrongdoing by employees. Thus, business and industry leaders, whether at the level of chief executive officer or in a compliance division, will find these cases helpful in designing effective ethics programs and codes of conduct. Hal Taback's essay on case use and his values checklist in Appendix A will help professionals in the field and educators in the classroom to focus their case discussion.

Engineering and science faculty also will find the text an excellent catalyst for heightening ethics awareness among students because it provides a set of cases that can assist the professor in integrating ethics discussion into a wide variety of courses. While helping to satisfy accreditation requirements, this approach enables the teacher to address ethical concerns throughout the course since the cases individually do not require significant time commitments. Such discussion on a regular basis will make a lasting impression on future engineers and scientists. On the other hand, if practitioners or faculty wish to make more extensive use of the cases, they can turn to Appendix B, which describes the highly successful Ethics Bowl developed at the Illinois Institute of Technology under the leadership of Dr. Robert F. Ladenson and described in this volume by Andrew Rehfeld, a graduate student in political science at the University of Chicago and the codirector of the 1997 Ethics Bowl.

ACKNOWLEDGMENTS

The editors would like to thank the many individuals who wrote cases for this project. The initials of the author of each case are given after the case title. Full names appear in the list of contributors. Manhattan College faculty members Drs. Bernard Harris (electrical engineering), Bahman Litkouhi (mechanical engineering), and Robert Mauro (electrical engineering) also were very helpful in working with students in their departments, and their cooperation is appreciated. Special thanks also go to Dr. Joseph Reynolds, Ms. Heidi W. Giovine, and Ms. Nancy Cave for their precise and insightful technical support.

John R. Wilcox,
Director, Center for Professional Ethics
Louis Theodore,
Professor of Chemical Engineering
Manhattan College

Introduction

CAN ETHICS BE TAUGHT?

Professionals are often skeptical about the value or practicality of discussing ethics in the workplace. When students hear that they are required to take an ethics course or if they opt for one as an elective in their schedules, they frequently wonder whether ethics can be taught. They share the skepticism of the practitioners about such discussion. Of course, both groups are usually thinking of ethics as instruction in goodness, and they are rightly skeptical, given their own wealth of experience with or knowledge of moral problems. They have seen enough already to know that you cannot change a person's way of doing things simply by teaching about correct behavior.

The teaching of ethics is not a challenge if ethics is understood *only* as a philosophical system. Sharon Deloz Parks notes that teaching ethics is important, but "if we are concerned with the teaching of ethics understood as the practice of accountability to a profession vital to the common good, the underlying and more profound challenge before all professional schools [and other organizations] is located in the question, How do we foster the formation of leadership characterized, in part, by practice of moral courage?" (Parks, 1993, 181).

Moral courage requires knowing *and* acting. College and university educators, as well as those charged with ethics training in the private sector, develop a sense of uneasiness when topics such as "fostering leadership formation," "moral courage," or "knowing *and* acting on that knowledge" are mentioned. Such terms resurrect images of theological indoctrination, Sunday school recitations, or pulpit sermonizing. These images contrast sharply with what the present-day professor envisions as the groves of academic freedom and dispassionate analysis. Perhaps out of fear of disrespecting the dignity of students and devaluing their critical reasoning powers or their ability to understand where the truth lies, faculty will take a dim view of academic goals that go beyond those strictly cognitive. The consequence of such values among the professoriate is the further erosion of a moral commons where an agreed-upon set of values and beliefs allows for discourse on ethics. Of course, the erosion has continued steadily from the inception of the Enlightenment Project in the seventeenth century until the present day wherever industrialized and postindustrialized societies have been subject to rapid cultural, economic, political, and technological change. It is not simply an erosion in the realm of higher education. Practitioners in the engineering and scientific communities experience the same erosion of the moral commons taking place in society as a whole.

The editors are certainly in agreement with their colleagues in higher education and those who do ethics training in the private sector, that individuals are not to be manipulated or indoctrinated. However, they are also convinced not only that students and other participants in ethics analysis must have a body of knowledge, but that they have a responsibility for the civic life of American society. Such responsibility requires leadership, moral courage, and action. Of course, none of these characteristics can be demanded or forced, only elicited. That is the great, yet delicate challenge facing the professoriate and all those charged with ethics training in other sectors. Eliciting a sense of civic responsibility as a goal of ethics analysis can be realized only as a derivative of cognitive processes and not as a direct goal. In sum, the formation of personal character and the practice of virtue are not to be subject to external control and the diminution of individual freedom through manipulation or indoctrination.

The moral life, Parks claims, "depends on the quality of the images held at the heart's core" (Parks, 182), and understanding those images is a key to ethics education. Actions, says Parks, flow from our meaning making: "what we perceive to be ultimately true and dependable in the most comprehensive dimensions we can conceive" (Parks, 183). Thus, images of success, perceptions of the world, anticipation of important issues in the future, all constitute one's worldview and have great impact on decision making. An individual's worldview, understood as one's basic assumptions about life or the "way the world works," determines the sense of personal accountability in that these assumptions give rise to a set of values and a valuing process whereby one makes choices concerning what is highly prized or important (Morrill, 1980, 63 ff.).

Both the set of values and the valuing process itself give rise to the moral choices one makes on a daily basis. When the term "moral" is used, it is frequently equated with ethics. In this introduction, however, the terms are defined somewhat differently. Choices have a "moral" quality insofar as they are decisions that have an impact on the lives and decisions of other persons. Increasingly in this period of history, choices are also judged to have a moral quality if they have a impact on other species or the environment itself. The study of ethics is a philosophical discipline that draws on human reason and analysis to assess these moral choices, and uses the rich tradition of ethical reflection in Western and related philosophical systems.

The editors believe that the case study method is a valuable way to take seriously Parks's response to the question "Can ethics be taught?" They also consider the method an important tool in investigating the relationship among assumptions, values, and the moral life, as well as ethical reflection on those three aspects of life. The editors are convinced that the case study method is one of the most useful way of teaching ethics and of achieving the goals of ethics education outlined by the Hastings Center (*The Teaching of Ethics in Higher Education*, 1980, 47–52):

1.   *Stimulating the moral imagination.* The concreteness of the case study appeals very much to the learning style of most people. While a certain amount of ambiguity is essential to evoke interest and discussion, it is also a stimulus to enlivening the knowledge. Hopefully, the participant will begin to appreciate the moral complexity of a situation that in the past might have been thought of only as a technical or managerial problem. Practice in the art of case discussion has the larger intent of leading the individual to bring an ethical frame of reference to bear on the variety of problems faced in the discipline studied. Stimulating the moral imagination is similar to putting on a pair of glasses that are tinted. The result is that the world is seen through that tint. As a consequence of the case study method, the editors and authors of the cases hope that each individual will see his or her field of study through the interpretive glasses of engineering and environmental ethics. He or she would then routinely ask: "What is the moral issue here?"

2.   *Recognizing ethical issues.* The case analyst should not be content with a good "imagination." The further challenge is the recognition of specific moral problems and how they differ from one another in terms of immediacy or urgency. Concreteness is an important asset of the case study and clearly assists in achieving this second goal. Comparing and contrasting a variety of cases through discussion is essential to recognition and leads to achievement of the next goal.

3.   *Developing analytical skills.* Differentiation, comparison, contrasts—all of these must be related to an enhanced ability to solve the problem. To achieve this goal, the student of ethics is taught to bring the skills developed in his or her major field of study to bear on the ambiguous situation, the moral dilemma, or the competing values that must be addressed. Analytic skills are best honed through the use of examples or cases. The technical ability to analyze all dimensions of an environmental spill will have an impact on how the moral aspect of the problem is understood in terms of resolving the problem. Of course, ethical systems that emphasize the importance of consequences, the obligations inherent in a duty-based ethic, as well as theories of justice or virtue will enhance the ability to use technical or discipline-based analytic skills in resolving the problem. Knowing, however, is related to acting. This leads to the fourth goal.

4. *Eliciting a sense of moral obligation and personal responsibility.* Much has already been said about the importance of this goal. However, it should be clear that a sense of moral obligation does not mean there is one set of absolute answers. Dictating a solution is quite different from an internalization process whereby the individual commits himself or herself to be a "seeker," one who takes personal responsibility for addressing and resolving the moral problems facing engineers or scientists. Both professions constitute the "guardians of the system" in the technical community. They are the first line of response to the problems and dilemmas facing the professions as such. To point to the Environmental Protection Agency, the Occupational Safety and Health Administration, the Federal Bureau of Investigation, congressional formulators of public policy, or other sovereign countries as the parties responsible for resolving acute problems is to abnegate one's moral responsibility as a professional person. To say this is not to dictate solutions, but to alert individuals to their personal responsibility for the integrity of the respective field. Eliciting a sense of responsibility depends on an assessment of the assumptions or "images at the core of one's heart." Assessment of ethical systems or normative frames of reference must be connected to the actual assumptions or images that constitute a person's worldview. Challenging the individual to examine that worldview in relation to a case and ethical systems is the first step in joining doing to knowing. Closely related to the achievement of this goal is the following one.

5. *Tolerating—and resisting—disagreement and ambiguity.* An essential component of case discussion is the willingness to listen carefully to the points of view held by others. Cases, by their nature, are ambiguous. They are bare-boned affairs meant more to be provocative than to lead to a clear-cut jury decision. The purpose of the case is to stimulate discussion and learning among individuals. As a result, there will be much disagreement surrounding the ethical issues in the case and the best option for resolving it. Toleration does not mean "putting up with people with whom I disagree." Respect for the inherent dignity of the person and a willingness to understand not only another position, but also a person's reasons for or interest in that point of view, should be part of the case discussion. Toleration does not mean all opinions must be of equal value and worth. It is true that respect for and listening to another person's argument may lead one to change a position. However, a careful description and discussion of the other person's position may also lead to a greater conviction that one's own position is correct. What is clearly of central concern is the belief that the free flow of ideas and carefully wrought arguments, presented from all sides without fear of control manipulation, threat, or disdain, is at the core of human understanding and development. This hallowed concept of academic freedom is the catalyst that allows human communities to be committed to the search for truth, without at the same time declaring absolute possession of the truth.

Ethics Must Be Taught!: Idealism and Pragmatism in Engineering and Environmental Ethics

Scenarios are, for the most part, designed to reflect ambiguity in work situations. The ethicist hopes to get his or her hands dirty, dealing with the bottom-line motives of survival, competitiveness, and profitability as well as the mixed motives of self-interest, respect for the rights of others, and altruism. Obtaining an ethical solution to a difficult moral problem or dilemma is based on much more than choosing the correct ethical framework with its normative frame of reference. One must also be ready to examine fundamental assumptions and the values to which the assumptions give rise. Stephen L. Carter has made this point recently in a discussion of "integrity."

1. *Honesty in relation to integrity.* Carter explores integrity in relation to the value society places on honesty. On this subject, one of the best-known and most popular ethics books of the last few decades is Sissela Bok's *Lying: Moral Choice in Public and Private Life* (Bok, 1978). Without taking away from the merits of *Lying*, Stephen L. Carter notes: "Plainly, one cannot have integrity without being honest (although, as we shall see, the matter gets complicated), but one can certainly be honest and yet have little integrity" (Carter, February 1996, 74). Honesty is far easier to practice than the tough work of figuring out what it takes to have integrity in a situation. Integrity requires a high degree of moral reflectiveness. Honesty may result in harm to another person. Furthermore, "if forthrightness is not preceded by discernment, it may result in the expression of an incorrect moral judgment" (Carter, 75). The racist may be transparently honest, Carter declares, but he certainly lacks integrity because his beliefs, deeply held as they might be, are wrong. He has not engaged in the hard work of examining his fundamental assumptions, values, beliefs.

2. *Personal integrity without public responsibility?* It would appear that one cannot have integrity without responsibility, since any consideration of integrity addresses the effects of our conduct on other people. In our work life and our community life, we have public responsibilities for our clients and fellow citizens. That is the nature of public life. It demands civic virtue of us. In this light, consider an example supplied by Carter:

> Having been taught all his life that women are not as smart as men, a manager gives the women on his staff less-challenging assignments than he gives the men. He does this, he believes, for their own benefit: he does not want them to fail, and he believes that they will if he gives them tougher assignments. Moreover, when one of the women on his staff does poor work, he does not berate her as harshly as he would a man, because he expects nothing more. And he claims to be acting with integrity because he is acting according to his own deepest beliefs.
>
> The manager has the most basic test of integrity. The question is not whether his actions are consistent with what he most deeply believes but whether he has done the hard work of discerning whether what he most deeply believes is right. The manager has not taken this harder step.
>
> Moreover, even within the universe that the manager has constructed for himself, he is not acting with integrity. Although he is obviously wrong to think that the women on his staff are not as good as the men, even were he right, that would not justify applying different standards to their work. By so doing he betrays both his obligation to the institution that employs him and his duty as a manager to evaluate his employees.   (Carter, 75–76)

Carter's reasoning concerning the hard work leading to integrity must be applied to the cases in this text. Answers to problems or dilemmas are not easily arrived at and require a willingness to examine one's fundamental assumptions, values, and beliefs. The theme of integrity plays itself out in a somewhat different fashion in Andrew Stark's provocative essay "What's the Matter with Business Ethics?" (Stark, May–June 1993). He notes that managers are not getting the needed help from business ethicists in addressing two types of ethical challenges:

> first, identifying ethical courses of action in difficult gray-area situations and second, navigating those situations where the right course is clear, but real-world competitive and institutional pressures lead even well-intentioned managers astray (Stark, 38).

Much as Carter faults those who opt for the easy road of "honesty," Stark faults business, and by extension engineering or environmental ethicists, for offering "a kind of ethical absolutism that avoids many of the difficult [and most interesting] questions." Such absolutism devalues the bottom-line interest and marketplace success. Ethicists of the absolutist persuasion would rather see the corporation sink than compromise idealism. Stark takes as the starting point the existence of the corporation and managers who "still lack solutions for the basic problem of how to balance ethical demands and economic realities when they do in fact conflict" (Stark, 43). The litmus test for all applied ethics, then, is whether it is an ethics of practice, a "dirty-hands ethics." A practitioner of such ethics "must help managers do the arduous, conceptual balancing required in difficult cases where every alternative has both moral and financial costs" (Stark, 43). Furthermore, these ethicists must address the complexity of personal motivations. "The fact is, most people's motives are a confusing mix of self-interest, altruism, and other influences" (Stark, 43).

The new business ethic—and, by extension, engineering and environmental ethics—may be identified in the following way:

> Moderation, pragmatism, minimalism: these are new words for business ethicists. In each of these new approaches, what is important is . . . the commitment to converse with real managers in a language relevant to the world they inhabit and the problems they face. That is an understanding of business ethics worthy of managers' attention (Stark, 48).

Ethics Education: The Practical Decisions

Stark's identification of a new business ethic has important ramifications for those intent on teaching engineering and environmental ethics. The engineer and scientist called to integrity in his or her search for a relevant ethic will no less demand realism than the business person seeking realistic answers in business ethics. There are no easy answers in the work life of those called to be engineers or scientists. A critical tool in arriving at realistic answers certainly can be found in the case study method. Case discussion provides an effective, fast-paced, learner-centered experience. Furthermore, this method and the case content engage the participants. The case method is valuable for four reasons:

1. Cases used are ambiguous and, therefore, provocative.

2. The cases contain issues with which the learner can identify or has experienced, or they can lead to topics about which the participant has a strong opinion.

3. Cases are sufficiently controversial to generate divergent opinions and therefore engage the learners in a conversation among themselves.

4. The case discussion decenters the teacher as the authority figure, transforming the educator's role into that of questioner, coach, listener, critic, resource person, who demonstrates total involvement in and works closely with the class.

Two Important Questions

Despite the great teaching advantage that comes with case use, there are two important questions that case discussants must keep in mind when they assess the ethical problem:

1. *Who are the guardians of the system?* This question addresses the issue of who, among engineering or science professionals, is responsible for the ethical standards in the organization. If professionals point the finger at senior management, the legal department, the Environmental Protection Agency, or the Department of Justice, they have indeed misunderstood the nature of a professional calling. The first line of defense is the willingness of professionals themselves to maintain and enhance the integrity of the engineering or scientific profession through their own personal adherence to the highest standards of conduct and to assume responsibility for commitment to these standards within the companies where they work. Moreover, ethics is a positive task, not a list of dos and don'ts. To achieve excellence in one's work presumes a commitment to the client's contract, public safety, and environmental integrity, among several factors that are all too often thought of as "management" issues. They are, in reality, the ethical standards of the work itself. Thus, the ethical engineer or scientist is the one who identifies with the profession and all that is involved in the work assigned or contracted.

2. *Who gives support to the guardians of the system?* This second issue goes to the heart of the assessment problem, but also has an impact on the first issue. Unless the organization backs those who assume positive responsibility for the ethical tenor of the group, very little will change. Why would someone risk ostracism or retaliation by confronting a person engaging in unethical behavior or illegal behavior if there is no institutional support for the one assuming responsibility?

Effective guardianship is facilitated if

1. There are clear-cut standards of behavior and high expectations of the membership.

2. The standards are brought to the attention of the members through a well-developed training program.

3. The standards are taken seriously by the senior leadership team of the firm. They must demonstrate that seriousness by taking an active role in the training, without, at the same time, creating a chilly climate stifling discussion and participation in the training. The ethics program must be seen not as frosting on the cake or as a value added on to forestall legal problems through better compliance. The CEO needs to demonstrate a commitment to the values and principles that drive the business. Ethics training is no add-on. Ethics is what *drives* the organization: trust, integrity, fidelity to the client.

4. It is evident that the leadership "walks the talk" in all aspects of its decision making and actions.

5. There are mechanisms in place to address the concerns of the members, mechanisms such as an ombudsperson, a hotline.

6. Those who adversely affect the integrity of the business are effectively and fairly disciplined.

Another way of addressing the question of who supports the guardians is to emphasize the importance of organizational or corporate culture. A positive response to the six points just raised has a great impact on the culture of the organization. Unless there is what is sometimes called a "thick" culture, wherein respect for and adherence to guardianship and the tenets of integrity and trust are palpable, individually ethical persons can do very little to raise the moral climate. An organization is more than the sum total of the individuals who constitute the membership. The attitudes conveyed, values expressed, and ways of doing business in an organization profoundly affect the perceptions of the members therein and set the tone of the company. Having a positive impact on culture is a great challenge and not easily achieved. Culture is so subtle that one often does not even realize or understand its dimensions until a significantly different culture is experienced.

An organization will not have effective guardianship of the system unless there is a concerted attempt to create, enhance, or reinforce a culture where values and ethics are clear and fully supported. There is little doubt, however, that the twin issues of guardianship and culture are much more difficult to address than the institutionalization of the ethics program itself.

BIBLIOGRAPHY

Bok, Sissela. 1978. *Lying: Moral Choice in Public and Private Life*. New York: Pantheon.

Carter, Stephen L. February 1996. "The Insufficiency of Honesty." *Atlantic Monthly*, 74–76.

Goldberg, Michael. 1993. *Against the Grain: New Approaches to Professional Ethics*. Valley Forge, Pa.: Trinity Press International.

Kaplan, Jeffrey M., Joseph E. Murphy, and Winthrop M. Swenson. 1993. *Compliance Programs and the Corporate Sentencing Guidelines.* Deerfield, Ill.: Clark Boardman Callaghan.

Kimball, Bruce A. April 1995. *The Emergence of Case Method Teaching, 1870s–1990s: A Search for Legitimate Pedagogy.* Bloomington, Ind.: The Poynter Center, Indiana University.

Morrill, Richard. 1980. *Teaching Values in College.* San Francisco: Jossey Bass Series in Higher Education.

Parks, Sharon Daloz. 1993. "Professional Ethics, Moral Courage, and the Limits of Personal Virtue." In *Can Virtue Be Taught*, edited by Barbara Darling-Smith. Notre Dame, Ind.: University of Notre Dame Press, 175–193.

Pritchard, Michael S. N.d. *Introduction to Teaching Engineering Ethics: A Case Study Approach.* Kalamazoo, Mich.: Center for the Study of Ethics in Society, Western Michigan University.

Stark, Andrew. May–June 1993. "What's the Matter with Business Ethics?" *Harvard Business Review*, 38–48.

The Teaching of Ethics in Higher Education. 1980. Hastings-on-Hudson, N.Y.: The Hastings Center.

Contributors

HA/NN/RV	Herry Alex, Nghia Nguyen, Robert Varghese
RA/SAN	Raged Abbassi/Sameer Al-Naji
MEB	Moheb Beshara
CC	Chris Chilcott
EC	Elena Capone
KJC	Keith J. Colacioppo
RC	Richard Carbonaro
RTC	Robert T. Ciotti
SC	Sean Culliney
BD	Bella Devito
GD	Gianni Del Duca
RD	Roberto Diaz
SE	Sherif Elshafei
PF	Paul Fernandez
KF	Karen Fox
SF	Saul Fergerson
YF	Yolanda Foley

AG	Antonio Gulino
BG	Belinda Gunn
JG	Jennifer Gagnon
SSH	Steve S. Harris
BRH	Brian Hawkins
CH	Christine Hellwege
BH	Bruce Howie
TJ	Travis John
JL	James Lippencott
JMSL	James Love
SM	Sameet Master
CM	Carlos Miranda
JM	James Morrissey
RM	Rosario Moschitto
NN	Nghia Nguyen
ROBP	Robert Pettenato
RP	Richard Phillips
CP	Christopher Pilek
CR	Christian Racic
GR	Greg Recine
AR	Andrew Rehfeld
MR	Marybeth Reynolds
NS	Nikki Sanders
JS	Jennifer Sang
AHS	Altaf H. Sarker
RS	Richard Shaw
AS	Andrew Shelofsky
JNS	Jennifer Spero
HT	Hal Taback
DT	Deanshi Trivedi
GT	Gildaro Tseng
ZT	Zhenyu Tang
GV	Gerard Vincitore
PW/CC	Patrick Whelan/Ciro Cuono
AW	Andrew Witt
MZ	Michael Zeolla

Part *I*

Environmental Ethics

1

Air, Solid Waste, and Water Pollution

A. ETHICAL CONFLICTS INVOLVING INTEGRITY OF DATA/REPORTS

1. Air Pollution

> **1.1** **Case Title:** I've Had Enough (ec)
> **Case Type:** Air Pollution

Fact Pattern

Stacy is a junior majoring in environmental engineering at Manhattan College and has decided to look for a summer internship. Given her interest in air quality, she sends her resume to a variety of energy plants in hopes of obtaining a position.

Frank, operation manager at NRGRUS, a power plant in Connecticut, cannot get over how far behind he is in his work. Fortunately, he receives a memo stating the company has received a handful of resumes from students interested in summer internships. As luck would have it, Frank decides to give Stacy a call and see if she is interested. Stacy is ecstatic and accepts the position.

The spring semester comes to an end, and Stacy begins her job. The first week is introductory, but then she is bombarded with work. What starts out as an internship ends up being "slave" work. Stacy has to test for concentrations of SO (oxides of sulphur), compare it with past data, and perform an analysis with all the data. She sometimes stays at the office until 10 or 11 P.M.

One week Stacy notices all her data are relatively the same. Therefore, in order to give herself a break, she decides to make up data for the next couple of days, based on the trends she has seen. A week later, Frank approaches Stacy and reports that all the analyses look good, except for a few days. Unfortunately, those are the days she made up the data. Frank asks her to redo the analyses and get back to him.

Stacy does not know what to do. She is not sure whether to tell him the truth or to redo the analysis. However, she fears he would think her incompetent for making such a drastic error of judgment as falsifying data. Stacy is truly in a bind.

Questions for Discussion:

1. What are the facts in this case?

2. What are the ethical issues in this case?

3. What will Stacy do?

4. What should Stacy do?

5. Is Frank right in giving her so much work to do?

1.2   Case Title: No More Late Nights (gd)
Case Type: Air Pollution

Fact Pattern

Joshua works in the environmental division of the ABC Asbestos Testing Company. His position is senior laboratory analyst. He is in charge of analyzing transmission electron microscope (TEM) airborne asbestos samples. In the past few weeks a client has been sending him samples that have been failing, or coming up positive for asbestos. When the results of the samples get back to the client, that client has to clean up the site again and take new samples. He then has to send the new samples back to the lab for reanalysis. The new samples have been coming to the lab at approximately 10:30 P.M. for immediate turnaround; Joshua is paged by the client to return at night to the lab and analyze the new samples. It takes a minimum of three hours to prepare and read the TEM samples, so Joshua goes home at around one in the morning.

After two weeks, Joshua's sleep patterns are disrupted. He cannot take much more of this, and decides to pass the next set of samples so that he can catch up on some sleep. The next day the client calls him on the phone.

"Joshua, please."

"Speaking," mutters Joshua reluctantly because he recognizes the voice on the other end of the line.

"Guess what?" says the client. "I have another set of TEMs that need immediate turnaround time because this building has to reopen tomorrow."

"Sure thing. Those samples will be done in a jiffy," exclaims Joshua.

"I'm surprised to see that you are so enthusiastic about reading my samples, Josh. I figured that you would be fed up by now; you've had to come back at night seven of the past ten days to analyze my resamples," says the client.

"When will they be here?"

"They should be there around six o'clock," says the client confidently.

"Well, all right, then . . . talk to you later," says Joshua.

"Damn it!" yells Joshua after slamming down the receiver. "I was planning to leave early today. It's a nice day outside, and I wanted to work on my car so I can get it ready for the winter."

Then he thinks to himself, "I guess I'll have to work on my car another day."

Six-thirty rolls around and the samples haven't arrived yet. Finally at seven o'clock the client brings the samples to Joshua.

"Sorry, Josh, the field technicians didn't finish sampling until just now. Here you go," the client says as he rushes to his car to get home for the big game on TV.

At this point Joshua tells himself, "Forget it. I was going to analyze the samples and come back tonight if necessary, but I'm fed up. No more late nights!"

Joshua decides to forgo reading the samples this time. He calls the client from his couch at home *three hours later* to disguise the fact that he didn't really analyze the samples.

"Hey, aren't I lucky today! The samples are negative for asbestos," says Joshua without even a quiver in his voice.

The client answers, "Great. Talk to you tomorrow."

Questions for Discussion:

1.   What are the facts in this case?

2. Why does Joshua want to pass the next set of samples that come in from the client?

3. Why does he pass the late set of samples for the client?

4. What risks does Joshua take by not analyzing the samples?

5. What will happen to the people in the building if the samples *are* actually positive for asbestos?

6. What will happen to Joshua if the client finds out that he didn't read the samples?

7. What can the client do to Joshua legally if he finds out?

8. Do you think Joshua will ever pass another set of samples without reading them again?

1.3   Case Title: Government Knows Best (as)
       Case Type: Air Pollution

Fact Pattern

During the 1996 vice-presidential debates, one of the candidates said that 10 percent of the automobiles cause 100 percent of the air pollution. While that is factually not the case, automobiles *do* contribute significantly to the problems with air pollution. The Clean Air Act of 1990 established a nationwide map of the country's air quality ratings and provided the Environmental Protection Agency (EPA) with guidelines for changing a region's air quality rating. Many areas of California, primarily around Los Angeles, are deemed to have "serious" air pollution problems, a consequence of its large populace, the quantity of vehicles on the road, and other factors, such as industrial pollution and the typical weather pattern of the area, which does not allow pollution to dissipate away from the city.

New York City and many of the surrounding suburbs are also considered to be "serious" air pollution regions. To the northwest of the city, several southern Orange County communities are in the "serious" air pollution category. A "serious" designation requires mandatory automobile emissions testing and pollution control measures for industry; in addition, larger businesses in the region must create ride-sharing opportunities for their employees, to reduce the number of vehicles on the road as people travel to and from work. For the past several years, the Orange County government and the New York State Department of Environmental Conservation (DEC) have requested that the "serious" designation be lowered to "moderate" for southern Orange County, citing that resulting air pollution studies in that designation were not performed correctly. However, the EPA has not modified its stance.

Furthermore, recent testing performed by the DEC in *northern* Orange County as well as Dutchess and Putnam counties revealed that there were two instances of high ozone levels in July. This brought the total number of instances of high ozone recorded to four in the past three years. Under federal law enforced by the EPA, a region can have no more than three instances of high ozone in a three-year period. These recent high-ozone levels place northern—*and* southern—Orange County in the expanding "serious" air pollution area. The EPA is currently reviewing the data and will decide shortly whether to redesignate the three-county area.

Many people inside and outside of government feel that the studies performed by the EPA are flawed, that the EPA guidelines are inflexible, and that the restrictions placed on citizens living in and industries operating in "serious" air pollution zones are too demanding. On the other hand, there are many individuals who look at the results of the Clean Air Act and feel that strict enforcement of the guidelines is the best way, perhaps the only way, to maintain a livable environment today and for the future.

Questions for Discussion:

1. What are the facts in this case?

2. What are the benefits to southern Orange County if it wins litigation concerning the "serious" air pollution rating?

3. What are the benefits to the DEC if it wins litigation concerning the "serious" air pollution rating in southern Orange County?

4.   What are the benefits to residents of Orange County if the air pollution rating is maintained as "moderate"?

5.   What are the benefits to industry within Orange County if the future air pollution rating can be maintained as "moderate"?

6.   What are the risks to the EPA if the air pollution rating is maintained as "moderate"?

7.   What are the risks to the residents of Orange County if the air pollution rating is maintained as "moderate"?

8.   What are the risks to industry within Orange County if the air pollution rating is maintained as "moderate"?

2. *Solid Waste Pollution*

2.1 **Case Title:** What They Don't Know Won't
Hurt Them (pf)
Case Type: Solid Waste Pollution

Fact Pattern

Jake is a geologist working on a current site evaluation and report for a project that his company has been hired as a subcontractor to conduct. The scope of the project involves transforming a solid waste dump into a park through habitat creation and land redevelopment. This project has been deemed essential to the vitality of Jake's company if they are to stay in business and continue to subcontract to the much wealthier firm on other future projects.

Jake has been informed that his evaluation and report should be limited to the current state of the land as it applies to the goal of creating a park. While conducting his research, Jake comes across a report on the land that was written fifteen years ago. Although Jake doesn't plan on using this report as a reference, he reads it out of curiosity and discovers that a section of the land belonging to the current project site was once accidentally contaminated with toxic waste by Jake's own firm. The toxic waste was promptly cleaned and the EPA designated the area to be safe. Jake informs his supervisor, Marie, of his findings and asks whether this information should be included in his report for the contracting firm.

"No, why would you want to include that information in your report?" Marie asks. "That happened fifteen years ago and was completely taken care of. Besides, your report is supposed to be limited to the current state of the site; isn't the land fine?"

"Yes, Marie, the EPA recently checked the land and was delighted by our plan to turn the site into a park."

"Well, then, you have your answer. If the land is fine, why would you want to cause any alarm that might upset the contracting firm? As you already know, Jake, our company needs the work that this firm is providing us, or we'll all be out of a job."

"You're right. I was just afraid that the firm might find out about the report later and wonder why we hadn't told them," Jake mentions.

"That is the only copy of the report we have, so just don't show them," Marie states.

"I guess you're right," mutters Jake.

"Trust me, do what I say and none of us will have to worry about looking for a job tomorrow," Marie says.

Questions for Discussion:

1.   What are the facts in this case?

2.   What is the problem Jake is facing?

3.   Is it an ethical problem? How so, or why not?

4.   What are the risks Jake and the company might encounter if he reveals the findings of the report?

5.   What are the risks Jake and the company might encounter if he *doesn't* reveal the findings of the report?

6.   Is Marie's advice beneficial or harmful to Jake?

7.   What final action do you believe Jake will take?

2.2   Case Title: Who Is Going to Notice? (jm)
Case Type: Solid Waste Pollution

Fact Pattern

Janet has been working at a material recovery facility (MRF) in Long Island for approximately five years. A brand-new MRF has just been opened in Yonkers—a technological wonder, primarily computer-run, with a minimal number of workers needed. Janet is asked to join the staff at this MRF, and she is thrilled.

Her first day there, Janet is given the usual tour of the facility, one-half of which deals with the recycling of paper, while the other recycles plastics, metals, and glass. The paper is transported by large trucks onto a moving platform where people sort the cardboard from the paper. Then the paper drops into a fifty-foot shaft, where it is crushed into packed piles. Likewise, on the plastics side, a rotating magnetic machine separates the metal recyclables from the plastics and glass. In assembly-line fashion, the workers separate number one from number two plastics (according to the numbers on the bottom of the containers), and each goes into a separate shaft.

These shafts culminate in separate bins aligned side by side on the bottom floor. Another moving platform lies beneath the bins to carry off the plastics to be crushed and packed in piles, just as for the paper. All of this is controlled via a panel on the top floor. There is a schematic diagram of all of the shafts, all the bins, and the moving platforms. When one of these bins is filled, a red light blinks on the control panel, signaling that the bin must be emptied.

"All you have to do is monitor this, Janet," her supervisor explains. "When the light blinks, push the green button next to the shaft to release the recyclables onto the platform so that they can be compacted. Do you have any questions?"

"Yes, what if two fill at the same time?" she asks.

"We don't expect that that could ever happen as long as you are monitoring the panel. If it *does* ever happen, the whole operation will be a disaster because you'll have to release them simultaneously, which will cause different types of recyclables to comingle on the platform below. Then we'd have to re-sort, hold up other bins . . . we'd fall drastically behind schedule."

A month passes, and Janet has had no problems operating the machinery. One day, however, she makes a huge mistake. Her manager is not in, so she takes an extended lunch with an old friend. When she returns, she realizes that two of the bins are filled. They are each filled with plastic recyclables: a number one and a number two. She decides to ask a coworker for advice.

"If I don't empty them both right now, I will have to explain how I let them fill. I may lose my job. What should I do?"

"Empty them at the same time. They're both plastics anyway, and who is going to notice?"

Janet listens to the advice of her coworker and pushes the buttons to empty both shafts.

Questions for Discussion:

1.   What are the facts in this case?

2.   Is Janet's friend right in her advice?

3.   What would you have done?

4.   Were there other options for Janet?

> **2.3   Case Title:** Explosion of Trouble (jns)
> **Case Type:** Solid Waste Pollution

Fact Pattern

Andrew is an engineer working in a pollution control company who has been asked by his boss to do an evaluation of a certain landfill they covered a few years ago. His main task is to run a check for methane gas at the landfill and on some of the blocks in the surrounding neighborhoods.

When Andrew goes to the landfill, he notices that most of the vents are emitting high concentrations of methane, and these concentrations are accumulating in pockets at some places along the fence-line of the landfill.

Andrew comes back from the landfill with a fifty-page set of data and hands it to Monica, a temporary employee, to present as a proposal to the city. Before leaving Monica's desk, Andrew comments in a brisk manner that even though the numbers seem high in some areas, he could not detect much of an odor. He also mentions under his breath that they would have to pay a high fine if the city does not like some of their numbers.

As Monica is reducing the data into readable tables, she notices that not only some but *most* of the numbers are extremely high around the adjacent neighborhoods.

Monica asks her coworker Judy, "Can't high levels of methane gas be harmful to people?"

Judy tells Monica, "Of course they can. Methane not only is harmful to health but also can cause explosions in the presence of fire or sparks."

Knowing that she should talk to Andrew about this first before saying anything else to Judy, Monica approaches Andrew.

She tells him, "These numbers for methane are very high, even in the neighborhood areas. I am going to have to report it that way."

Andrew tells Monica, "If you report it that way, we could lose a lot of money, time redoing the landfill cover, and even possibly our jobs. Look, no one is complaining, and there is no odor, so how is anyone going to find out, right?"

"Yeah, I guess so," replies Monica.

"It *is* okay, believe me. It doesn't matter if the numbers are a little off," retorts Andrew.

Questions for Discussion:

1.   What are the facts in this case?

2.   What effect will Judy's comments have on Monica's decision?

3.   What might happen if Monica does not report high numbers?

4.   What might happen if Monica *does* report high numbers?

5.   What final action do you think Monica will take?

3. *Water Pollution*

> **3.1 Case Title:** Muddy Waters (kjc)
> **Case Type:** Water Pollution

Fact Pattern

UMS is a large consulting firm that specializes in water treatment. Most of the work that it undertakes is for government agencies that must abide by strict equal opportunity laws as well as quality standards. UMS is therefore required to hire "disadvantaged" firms as subconsultants.

In the Northeast, where UMS is consulting on a large project for the Department of Environmental Protection (DEP), there are many available firms to subcontract to, but few that meet the equal opportunity requirements. TEC is a reliable firm that has done work for UMS in the past and is very proficient in terms of water treatment. Its methods are innovative and its record is unmatched in the field, but only 10 percent of its employees are minorities. SFN is a so-called disadvantaged firm bidding on the same job. Its record in the field is poor, it often leaves a site without reaching the quality standards imposed by the DEP, but it employs a 50 percent minority workforce—specifically, it seems to get government contracts despite its poor work quality. And, when the bids come in, SFN's bid is considerably lower than TEC's.

Questions for Discussion:

1.   What are the facts in this case?

2.   What is the problem UMS is faced with?

3.   Is UMS sacrificing water quality to meet some other irrelevant specification?

4.   Is it ethical for SFN to carry a 50 percent minority workforce to get more jobs even though its performance is poor?

5.   What if it had an excellent reputation?

6.   Who should UMS use?

> ### 3.2   Case Title: The Heavy Metal Dilemma (rc)
> ### Case Type: Water Pollution

Fact Pattern

A university that has a program in environmental engineering has been given a grant to do research in the area of heavy metals in lake sediments. The overseer of this research is a professor who is well respected not only at the university but also in the field. After this professor secures the grant, he puts to work various graduate and undergraduate students to carry out the research in the lab. The experiment involves examining the binding capacity of metals in sediments; more specifically, the ability of certain metals to form metal sulfides, thereby rendering the heavy metals harmless to any life in the sediment.

Paul, one of the ten student members of the research team, is selected to set up this experiment, which he does as follows: A long plastic cylinder is used as housing. The bottom of this cylinder is sealed, with the top left open. Sediment collected from the bottom of the lake is placed in the cylinder, and then water is placed on top of the sediment, to simulate lake conditions. A key parameter of this experiment is that the water be kept at a pH of 7, that being the normal pH of lakes. Over time, the water is sampled and analyzed for metals. The only maintenance on this experiment is that the overlying water must be refilled periodically because the volume decreases due to evaporation. This water is to be replaced with a special buffer made up in the lab and kept in two-liter flasks. Paul is to perform this experiment for four different cylinders so as to generate a reasonable amount of data from the experiment.

For about two months, Paul tends to the experiment. Occasionally he takes samples from the overlying water, and very often he has to refill the cylinders with buffer solution. There are some instances, however, when Paul cannot be in the lab to tend to his experiment; either he has to attend class or has to take a day off from work. Under these circumstances, one of the other members of the research team takes over Paul's duties.

One day when Paul is in the lab analyzing his samples of the overlying water from the sediment cylinders, he notices something peculiar. His samples read very low for the first two months of sampling, and then the concentration of metal shoots up suddenly. This happens on the same day of sampling for each of the four cores. The data seem to suggest someone tampering with his experiment; the concentrations he finds break all the rules of kinetics. Paul asks the other members of the research team what they think, but none of them knows what to make of the results. The only thing Paul can do is go to the professor and try to explain what went wrong with the experiment.

Paul does just that. In the meantime, a conversation takes place between Melinda and Jonas, two of the other lab researchers.

"It's a tough break for Paul if that experiment went bad," Jonas says to Melinda.

"Yeah, it really is," Melinda agrees.

"It would be a shame if someone really did tamper with his sediment cylinders. He put so much work into that experiment," Jonas continues.

"That *would* be terrible."

"Especially if it was one of his coworkers."

"Yeah, but I highly doubt anyone in here would be capable of something like that," Melinda responds.

"I'm not so sure about that," Jonas answers quickly. "Did you happen to see the jar containing the buffer solution that Paul uses to refill his cores?"

"Yeah, sure. I made up the buffer."

"Well, did you happen to see the *label* on the jar?" Jonas asks as he moves to the bench where the very jar sits.

"*Of course!* I made up the buffer and put the label on there *myself*," says Melinda.

"Why don't you look at the jar more closely?" Jonas replies.

As Jonas says this, Melinda looks at the jar and sees the source of the problem. Her face goes slightly pale as she answers him quickly, "Listen, I didn't realize that there happened to be two labels on the jar, one on either side. When I made the buffer solution, I removed one of the labels, and replaced it with a label that said buffer. I didn't know that there was a label on the other side of the flask that said that the flask contained 3 percent nitric acid. I also did not look into the jar to see if there was any liquid. How was I supposed to know? It's not my fault."

"Whether or not it's your fault is debatable," Jonas responds, "but what is definite is the fact that you now know what happened and you don't seem inclined to tell anyone. Take some responsibility for your actions. When Paul added the buffer with acid to the cylinders, the pH must have dropped down to around 1. At that pH, all of the metal sulfides were dissolved, freeing up all of the metal. That accounts for his results being the way they are."

Just as Jonas says this, Paul comes back into the lab.

"Did you get hold of him?" asks Jonas about Paul's attempt to see the professor.

"No," replies Paul, "but I'm going to go back in fifteen minutes and try to talk to him again."

Questions for Discussion:

1.   What are the facts in this case?

2. Is the confusion between the buffer and the acid Melinda's fault?

3. Should Melinda tell the professor what happened so that he doesn't blame Paul?

4. Is Melinda unethical in her decision to refrain from telling Paul what happened?

5. If Melinda does not tell the professor, should Jonas tell the professor?

> **3.3 Case Title:** High Tides (jg)
> **Case Type:** Water Pollution

Fact Pattern

This past summer has been extremely dry, with drought warnings as well as regulations for water use. The city's microbiology lab has been testing for total coliform and fecal coliform for thirty years; a team takes samples for these indicators at various spots along the harbor during high tide. Numbers have reached an all-time low this summer; and Linda Murphy, the microbiology lab manager, has reopened three beaches because fecal contamination has tested so low. She has been boasting about her wonderful management of the harbor and how her input concerning control factors has made such levels of water purity possible.

"Jack, look at these numbers! I have been managing this lab for fifteen years, and Sandy Bay has never been this low in contamination," Linda exclaims to a colleague.

"Yeah, Linda this is great now, but supposedly there is a hurricane coming in from the islands. All that rain will wash us out, and the numbers will definitely be huge. We've been lucky: We completely stopped the combined sewer overflow [CSO] abatements and leak detection with this drought," Jack replies with a worried look on his face.

"What exactly are you saying, Jack?! These numbers are a result of my management of the contamination and the wastewater plant upgrades that have occurred in the last year," replies Linda angrily.

"I totally agree that the upgrades have made a contribution, but with your education and experience you cannot seriously tell me that this drought is not the main reason why these numbers have remained so low," Jack says.

Linda goes home that night and watches the weather. Sure enough, the tail of that hurricane is about to hit the city. Despite their fears about any potential damage the storm could cause, the rest of the population is rejoicing over the prospect of being able to water their lawns soon, while Linda is biting her fingernails.

"What am I going to do? Jack is right. I prematurely stopped the CSO abatements and leak detection, and then proceeded to write five reports to Mr. Brown, my supervisor, crediting those programs. I just got that raise as well as the offer to speak at the national microbiology conference. I will completely lose my credibility," Linda thinks to herself.

The storm submerges the city for five straight days, stopping the sampling cruise. After it is finally over, the microbiology lab is flooded with calls from the health department wanting to know the status of the beaches. All Linda's employees are ready to hop on the boat early in the morning when the tides are high and the counts are most accurate. But where is Linda? Finally she shows up and puts off the cruise until as late as possible.

"Linda, we need to get out there. We're losing the tide. And the phone hasn't stopped ringing all day," Jack informs her.

"Soon. I need to fax a ton of information. Just give me an hour," replies Linda coolly.

"But the counts won't be as accurate if we delay, and we could be endangering the health of people if they swim in a contaminated area. The health department is breathing down our necks," Jack comments with a worried tone in his voice.

"I said, give me an hour," Linda replies sternly.

Questions for Discussion:

1.   What are the facts in this case?

2.   Why were the counts for fecal coliform and total coliform so low this summer?

3.   What is the problem that Linda is faced with?

4.   What is her reaction to the problem, and what will she do now?

5.   What does Jack alert her to?

6.   Why does Linda deter the sampling cruise?

7.   What would you do if you were put in Linda's position?

8.   What would you do if you were put in Jack's position?

B. ETHICAL CONFLICTS INVOLVING EMPLOYERS/MANAGERS

1. *Air Pollution*

> **1.1   Case Title:** Whose Side Are You On? (bh)
> **Case Type:** Air Pollution

Fact Pattern

Captain Smith is a nuclear engineer for the Air Force. He is a graduate of the Air Force Academy and has always been a "model soldier" by anyone's standards. Captain Smith has just been assigned his dream job, a top-level position at a former nuclear testing facility in the Southwest. It is his responsibility to measure ground and atmospheric radioactive contamination levels for the outlying area (approximately within a hundred-mile radius).

The Nuclear Regulatory Commission (NRC) has strict limits on the levels of radioactive dust that can be present in the atmosphere; Captain Smith must routinely report his findings to them. Due to the end of the cold war as well as the enforcement of United Nations test ban treaties, the base has not performed any live nuclear tests for almost two decades. Captain Smith has never recorded contamination levels above the "acceptable" range. However, the base has been operational since the early 1950s, so out of curiosity Captain Smith decides to look through the old records. What he discovers is that, as recently as a decade ago, the base was releasing as much as three times the amount of radioactive dust permitted by the NRC into the atmosphere. He also finds that a large civilian community 150 miles outside the base has experienced a great number of health problems that are characteristic of exposure to radioactive fallout. Believing these findings must involve an oversight or a mistake, Captain Smith decides to bring the records he has found before his commanding officer.

"Some of these claims are over thirty years old, Captain!" General Brown blasts. "The NRC has given us a clean bill of health. Whose side are you on anyway? It is your duty only to check *current* contamination levels, not to go snooping through documents that were printed before you were born."

"I thought that this matter might have been kept from your attention, and I felt it was my duty as the head nuclear engineer to report it to you directly."

"I appreciate your concern, Captain Smith. But do you think that if those claims had been at all relevant, we would've just ignored them? Of course not. Now, forget you even saw the reports and return to your assigned duties."

"Yes, thank you for your time and consideration, sir," says Captain Smith reluctantly.

"Very well, Captain, keep up the good work. Remember, what we do here is for the safety and well-being of all Americans. Dismissed."

Questions for Discussion:

1. What are the facts in this case and how do they challenge Captain Smith's sense of ethics?

2. What is the environmental problem that Captain Smith is facing?

3. What could be the possible consequences if Captain Smith does not report what he has found?

4. What could be the possible consequences if Captain Smith reports his findings to the NRC or to civilian authorities?

5. Is General Brown truly interested in the welfare of the community, or just his own? If so, why his urgency to dismiss the reports?

6. What do you think Captain Smith's final action will be?

1.2   Case Title: The High Life (jns)
Case Type: Air Pollution

Fact Pattern

Mary is an engineer in the air group in her company. Her job is to test for Volatile Organic Compounds (VOC) emissions and odors from hazardous waste sites, waste-water pollution plants, and sewer treatment plants. On a routine trip to one of the wastewater treatment facilities, for the purpose of taking odor surveys around the plant and on the rooftop, Mary is asked by her boss, John, to photograph some parts of the plant and the surrounding area. The photos will be included in a presentation and a report to show all of the progress made within the last few years.

As Mary is taking pictures on the rooftop, she notices that the view is beautiful. She begins to photograph some high-rises recently built right outside of the fence-line of the treatment plant.

In her excitement, Mary suddenly realizes that these buildings might have the potential to cause major problems for the plant. When the film is developed and Mary sees the pictures, she recognizes the potential problem: the stack VOC emissions. Six years ago, when the city agreed to a permit allowing the wastewater treatment plant a VOC emissions level from the stacks, it was based on the fact that the stacks were so high, there was no danger or harm to the people living or working in the area.

Mary decides to test the stacks for VOC emissions and also to run an odor survey with the people living in the new buildings adjacent to the plant. She finds that the VOC emissions are still in compliance with the permit of six years ago; however, they would probably be too high, out of compliance, based on today's regulations. Mary also discovers that the people living in the buildings have been complaining about sewage and fecal odors coming from the plant. Mary decides to act on these discoveries.

She says to John, "We have a problem with the stacks at the wastewater treatment plant I recently photographed. The stacks are now level with the apartments of some of the people who are living adjacent to the site."

John replies, "So, what's the problem?"

Mary continues, "The problem is that the VOC emissions would now unquestionably be out of compliance if the city were to run a test. Also, there are very bad odors coming from the stacks that the people have been complaining about."

"Hold it. Are the emissions still in compliance with the old regulations?" John asks.

"Well, yes, but . . . ," replies Mary.

"But *nothing*. Until and unless the city runs another test on that plant, we have nothing to worry about," retorts John.

"But what about the complaints from the neighbors?" asks Mary.

"We'll just use a masking agent in that area to cover up the odor, and that will do. There is no sense making a mountain out of a molehill. We are not going to start spending all this money for a problem that no one knows about. Just forget it and go back to work. I'll handle it," says John.

Questions for Discussion:

1.   What are the facts in this case?

2. What are the possible courses of action for Mary?

3. Mary tests the emissions and takes odor surveys to evaluate a potential problem. Should this really be her concern?

4. What are the risks if Mary remains silent and lets John "handle it"?

5. What can happen if Mary decides to act?

6. What final action do you think Mary will take?

1.3   Case Title: Too Good to Be True (ns)
Case Type: Air Pollution

Fact Pattern

Tamara works as an assistant engineer in a manufacturing plant. Over the past few months, the plant has been under strict watch by the state to maintain its discharge limits. The owner of the plant, Mr. Cox, has called an executive meeting to discuss a new product that will be made in the plant.

"Good afternoon, ladies and gentleman. As you know, I gathered you today to discuss a new product we will be manufacturing in the next two months," he says excitedly. "With this product we will net an extra $950,000 annually, plus we will gain publicity because the product was rated number one in the advance consumers' poll!"

"This sounds too good to be true, Mr. Cox. Are there any disadvantages to this product?" Tamara asks as all of the other executives look in Mr. Cox's direction, eager to hear his answer.

"Oh Tamara, there you go again, always thinking pessimistically. There are no *real* dangers in making this product—just a little risk, that's all," Mr. Cox replies, a little bothered by Tamara's inquisition.

"I don't intend to be the bad guy, but what do you mean, "There are no *real* dangers'? In my book, there can be *no* levels of danger, especially in this business."

As Mr. Cox looks around the room, he sees some executives nodding their heads in agreement with Tamara's statement.

"Well, Tamara, a by-product of the manufacturing process is a small amount of toxic fumes, nothing major."

"Nothing major? We are operating close to standards now, and we have the state breathing down our backs. Now you say we are going to produce a *small* amount of toxic fumes with our already-risky discharge? I don't think it's a good idea," she says angrily. Once again coworkers agree with Tamara, and Mr. Cox becomes furious.

"Well, did I forget to mention that with this annual net increase, employee wages will be raised?" As soon as he says that, many executives seem to begin to favor the product.

"That does not matter to me," says Tamara. "What about the public? Will they be informed that this plant will be producing harmful toxic chemicals?"

"Well, I would hate to spark controversy over this product, especially since the public loves it so much. I don't think *that* would be a good idea," Mr. Cox replies jokingly as other executives join in the laughter.

After the meeting, Tamara is angry and surprised that the owner of the plant could be so oblivious to public safety. Secretly, she likes the idea of a pay raise, but her conscience will not let her ignore the fact that toxic fumes generated with plant discharge could possibly harm or even kill people in the surrounding neighborhood. As she sits at her desk, she ponders the idea of calling the EPA to forewarn them about the product and her boss.

Questions for Discussion:

1. What are the facts in this case?

2. Why is public knowledge about the product important?

3. What will Tamara risk in calling the EPA?

4. What action could the EPA take against the manufacturing plant?

2. *Solid Waste Pollution*

2.1   Case Title: Don't Ask (ch)
     Case Type: Solid Waste Pollution

Fact Pattern

DAX is a multinational air conditioning manufacturer. They are presently facing potential loses due to low cash flow, high labor costs, and a decrease in customer confidence.

Bill, their new CEO, is hired to revitalize the company. He cuts middle management and suggests the hiring of MBAs within the next year. To generate cash, Bill orders Lou, their VP in real estate, to find a buyer for a large plot of underdeveloped, oceanfront DAX land in Florida.

Lou quickly receives a reasonable offer from a new developer of retirement villas. Although the developer is considering a number of available locations, Lou has definitely gotten his attention. The location would be perfect for elaborate walkways and recreational facilities. As a safety check, the developer has an environmental audit done on the area and is told that there is no potential problem with the property.

Before the sale is finalized, Lou is approached by Susan, a colleague who recently has heard rumors within the company that there is a significant amount of hazardous waste concealed within the location a few yards from the shore. After some difficulty, Susan confirms the rumor by inspecting the property herself. She goes on to explain to Lou that there are at least a dozen cracked barrels labeled DANGER BIOHAZARD seeping their contents into the ground beneath.

Lou approaches CEO Bill about the situation. He is sternly interrupted by Bill, who reminds Lou that this vital sale must be closed as soon as possible or the company will be facing possible bankruptcy. Lou decides to consult the company lawyer, who notifies him that Florida law does not require the disclosure of hazardous substances on commercial property as long as there is no misstatement about the condition of the property.

Questions for Discussion:

1. What are the facts in this case?

2. Is Bill simply doing what is best for the company? Is he acting ethically by ignoring the hazardous situation?

3. Should Lou inform the developer about the hazardous waste? What may happen to Lou if he does?

4. If Lou fails to inform the developer, does the lawyer's statement excuse him of wrongful doing? Why or why not?

5. Bill and Lou are not the only people who are aware of the situation. What about Susan? Where does her responsibility lie?

2.2   Case Title: Washing Away the Truth (ns)
Case Type: Solid Waste Pollution

Fact Pattern

Part One

Laura is the head lifeguard at the local beach. She likes to be the first person to arrive at work just to enjoy a little peace and quiet before the people arrive and things get hectic. As she stands at her tower and overlooks the ocean, she notices piles of washed-up garbage along the shore of the beach. The beach is scheduled to open in another hour, yet it is going to take an entire day to clean up this trash that extends almost the length of the beach. Laura becomes very upset as she walks into the storage closet to get that embarrassing sign, BEACH IS CLOSED DUE TO SOLID WASTE CONTAMINATION.

She goes inside her office to call all of the employees to let them know that they have yet another day off. In anger she calls the head administrator of the beach, Mrs. Kyra Morrell.

"Hello, Mrs. Morrell speaking."

"Hi, Kyra, this is Laura."

"Oh, hi. How's everything down at the shore? You know you guys are lucky, it's supposed to be a beautiful day today. Good for business."

"Well, I don't know about that. I'm going to have to close the beach again today."

"Oh no, don't tell me . . . trash," Kyra says as if this is a familiar problem.

"Yes, and lots of it."

"Well, what are you going to do?" Kyra asks, a little worried about Laura's tone.

"I'm going to call the board of health because things are starting to get out of hand," Laura says angrily.

"What are you, crazy? I don't think you want to get them involved."

"Why not?" Laura says, not understanding Kyra's reaction.

"The board of health is already swamped with work as is, and we'd just be put on a waiting list. Besides, if they get involved, the beach would have to be closed for at least a month for inspection. This could also lead to bad publicity for the beach, and I don't want that to happen."

"You mean to tell me that you don't think the board of health should know about this because it will arouse public concern? That's the reason I want to do this; it's called *public safety*. This is a very important issue that should be addressed. Besides, they know there's a problem; it says so on that stupid sign," replies Laura.

"Calm down. There's got to be another way to handle this. I think you're blowing this out of proportion," Kyra comments.

"*You* are not out here. You don't understand what it's like to turn away people and have to explain constantly why there are big garbage trucks on our shore. I have closed this beach five times this summer due to solid waste run-up. Each time there's been more trash accumulation than before, and it's occurring more frequently. If we clean it now, how do we know that it won't happen tomorrow or the next day? Besides, every time the beach is closed, we lose money. But safety, not money, is my major concern. I'm telling you, I'm fed up."

Questions for Discussion:

1. What are the facts in this case?

2. What are the personal dilemmas that Laura is experiencing?

3.   What obstacles are standing in her path?

4.   What options does Laura have?

5.   Are Laura's arguments for calling the board of health legitimate?

Part Two
After the conversation with Kyra, Laura decides to investigate further. She finds that this problem is not new to this season. It turns out that five years ago, this beach experienced the same problem. Kyra, who was the head administrator at that time too, was warned by the board of health that if it happened again, the beach would be considered "hazardous" and closed for inspection, which could take up to three to four years to complete. Laura feels strongly about the safety of the public, but if the board of health closes the beach, she will be out of a job. Why didn't Kyra tell her this in the beginning?

Questions for Discussion:

6.   What options does Laura have?

7.   What are the personal dilemmas that Laura is experiencing?

8.   What does Kyra's secrecy say about her business tactics?

> **2.3 Case Title:** Best-Kept Secret in Town (mz)
> **Case Type:** Solid Waste Pollution

Fact Pattern

Karen is an environmental management coordinator for EPA working on a current evaluation and report for the agency on the aftereffects of transforming solid waste landfills into productive designs for the community. The scope of the project is to take past data on a number of regional landfills that have been constructed and to coordinate it with data on the current usage (if any) of those properties. The purpose is to see if the government can justify the cost of spending so much money on remediating landfills.

Karen has been informed that the evaluation and report should be limited to the state of New Jersey. While conducting her research on past data, Karen comes across a remedial design report on a solid waste landfill that was written twenty years ago. Karen doesn't plan to use the report in her evaluation, but she reads it out of curiosity to learn how it might pertain to her project. In her reading, Karen discovers that a contractor employed by EPA had made some evaluation mistakes in the remedy chosen to protect the environment.

Karen informs a fellow coordinator, Peter, of her findings and asks whether she should inform the emergency response division about these findings.

"Yes, you *should* inform someone about your findings because recontamination of the site may cause damage to the surrounding environment," Peter exclaims. "Twenty years of uncertain protection may have caused serious problems in the area."

"You're right, but I was afraid that the agency might be just as liable as the contractor. I wonder why we've not told the public of the problem," Karen replies.

"You need to inform someone fast so that they can secure the site before something happens," says Peter.

"Right again," says Karen.

Questions for Discussion:

1.   What are the facts in this case?

2.   What is the problem Karen is facing?

3.   What risks is Karen taking if she does not tell anyone?

4.   What would take place if she reveals her findings?

5.   What final action do you believe Karen will take?

3. *Water Pollution*

3.1   Case Title: Blooms and Blame (ec)
Case Type: Water Pollution

Fact Pattern

Recently, people in the area of Kingsbridge, New Jersey, have been complaining of algal blooms in area waters. Reports have been issued to the area DEP, which has decided to investigate the Rock Island Treatment Plant, the largest plant in the area.

Susan, the operator at Rock Island, has been notified of these reports. Her analysis of sample data has revealed that nutrient levels have shown a slight increase in the past months. She therefore orders a thorough investigation of the plant.

John, an employee at the plant, reports to Susan that there appears to be an operational default causing the increased nutrient levels. Susan investigates the problem herself and finds that fixing the problem will cost the plant a significant amount of money.

"John, who discovered the problem within the plant?" Susan asks.

"Well, I did," responds John.

"We have to keep this between you and me. We will report that the plant is operating fine and propose that other treatment plants in the area be investigated."

"But we cannot do that," says John.

"Listen, fixing the plant will be very costly. Let us hope that the blooms do not continue."

"But how are we going to deal with it if they *do* continue?"

"I will worry about that," Susan responds. "For now, keep quiet or consider yourself unemployed."

Questions for Discussion:

1. What are the facts in this case?

2. What are the possible ramifications of this case?

3. Are Susan's actions ethical?

4. What do you think John will do?

5. What should John do?

3.2   Case Title: Let Them Worry about It (rc)
     Case Type: Water Pollution

Fact Pattern

John works at a wastewater treatment plant in Anytown, New York. As an entry-level engineer at the plant, he monitors the operating conditions of the chlorine contact basins that are presently in operation. The different parameters that John is to monitor include the flow of wastewater into the chlorine contact basins, the coliform bacteria count of the wastewater entering and leaving the tanks, and the chlorine residual of the effluent flow. The purpose of this monitoring is to ensure that the coliform bacteria count is within an acceptable level, one that conforms to the Environmental Protection Agency's maximum contaminant level (MCL). Coliform bacteria is not harmful to humans, but this type of bacteria is an indicator of the presence of other types that *will* cause harm. Thus far, John has not recorded any values for coliform bacteria that have exceeded the acceptable level, but, as an enterprising engineer, John discovers something else of interest.

Taking all of the data for the concentration of coliform bacteria from the contact basins from the past six months, John plots concentration versus time. Upon doing this, he notices a slight upward trend in the data, one that would never be noticeable when looking at the concentration of bacteria from day to day. The plant still is meeting the maximum contaminant level for coliform bacteria, but if this upward trend were to continue, in a few months the MCL would be reached. As soon as John is sure that his analysis of the coliform counts for the past few months is correct, his initial thought is to take his results to William, his supervisor. The first opportunity that John has to mention his findings to William is while his boss is alone in his office.

"William, do you have a minute?" asks John.

"Only if it's about the plans for the flow model you are working on," replies William. "The big guys upstairs want to know what kind of progress we are making."

"It's coming along well, but I have to do a bit more research on it. I wanted to talk to you about something else, though; it's about the coliform levels in the contact basin effluent."

"What about them?" asks William. "Don't tell me we've got a reading that is not in compliance! I mean, we didn't have a plant failure just now, did we?"

"No, nothing like that. It's just that I've noticed something about the data. The coliform bacteria levels have been slowly increasing over the past few months."

"How bad?"

"Well, if the increase continues at this rate, we would be out of compliance in a couple of months."

At this point, William sits back in his chair and does not say anything. He thinks hard about what John has just said. John then asks, "What are we going to do?"

"What do you mean, 'What are we going to do?'" responds William hesitantly. "We aren't going to do anything."

"Why not? I mean, I understand that we are in compliance now, but in two months or so . . . ."

"William, let me explain something to you about the way things work around here. The people upstairs—most specifically, my boss, Mr. Doe—have been hitting us with work nonstop for quite a while now. I can guarantee that if we show him the results you have compiled, he will ask me to assign someone to look into this problem, and he will still expect all of the other work to get done without increasing the number of personnel. We

are not out of compliance, so we don't have to tell anyone anything."

"William, I can look into it and still handle my other
work," replies John. "Let me look into it."

"I can't let you do that, I need you to finish up your
part of this flow model. It's more important right now,"
William answers.

"What happens when we start to approach coliform
levels out of compliance?" asks John.

"We let them worry about it upstairs, and we tell them
that we could have done something about it if we
weren't so overworked. It will be their fault, not ours."

Questions for Discussion:

1. What are the facts in this case?

2. Is the decision by William *not* to address the problem
an unethical one?

3. Is William justified in his decision?

4. Should John approach someone else about his findings? Should he approach someone of equal or greater
authority than William?

5. Should John look into this potential problem further
if it does not interfere with his regular work?

3.3   Case Title: Campaign Trail (jg)
Case Type: Water Pollution

Fact Pattern

It's election time again, and this year's campaign has been a heated one. Up for reelection is Mayor Martha Vineyard. Besides a number of other claims, she has been promoting herself as the environmentally conscious candidate. She has repeatedly pointed to the island's improved harbor and remarked about the number of permanent beach closures that were recently reversed.

"With Mayor Vineyard in office, you went to a clean beach this past summer. My team got rid of the floatable problem," she boasts. "You and your families were able to eat that lobster without worrying about pathogen contamination and to go clamming Sunday morning. I am the candidate who cares about the future of this island environmentally, and I proved it with the reopening of Beach A after ten years of contamination problems and Beach B after fifteen years."

The truth is that there have been major improvements in the water quality but not because of Mayor Vineyard *specifically*, if at all. The island has been working for thirty years to correct this problem. Federal regulation prompted the abatement of many combined sewer overflows as well as illegal sewer connections. The island improved the wastewater treatment facilities to stop primary effluent escaping the system. All of these continued while Mayor Vineyard was in office but the relevant legislation was passed by prior administrations. In fact Mayor Vineyard pulled some of the funding for research and gave it to her limo service.

Linda is an environmental engineer who works for the island's Department of Environmental Protection on the pathogen contamination project, which she has led for the past ten years. She is not really sure who to vote for, but she knows that Mayor Vineyard's boasts are not the whole truth. Two weeks before election time she is contacted by Mayor Vineyard's public relations director.

"Hey, Linda, we need you to do us a bit of a favor down here. Bill C., the newspaper reporter, would like to talk to you about the improvements in the harbor and the mayor's role in them."

"Well, the measures to effect these changes were approved before the mayor came into office, and I'm not sure she's really had a role in the improved water quality of the harbor," Linda states quite efficiently.

"Linda, the mayor's office is asking a favor from you," the PR director answers, his voice a bit strained. "The fact remains that those beaches opened while she was in office, so she's going to take the credit. If you cannot help us out with this matter, we will remember that when the election is over. *You* should remember you are still an employee of the island, and Mayor Vineyard is up in the polls!"

Questions for Discussion:

1. What are the facts in this case?

2. What is Mayor Vineyard implying in her campaign?

3. What were the harbor's water quality improvements?

4. What is the truth behind the improvements? Did Mayor Vineyard play an important role?

5. What is the public relations director asking of Linda?

6. What is the implication if Linda does not support the mayor's position?

7. Is what the public relations director asks of Linda an ethical task, if we disregard his implied threat?

8. What should Linda do? Come up with more than one option.

9. Determine Linda's *best* option and develop her course of action.

C. ETHICAL CONFLICTS INVOLVING PERSONAL INITIATIVE

1. *Air Pollution*

1.1 **Case Title:** Drive-by Polluting (sf)
Case Type: Air Pollution

Fact Pattern

John, a financially strapped engineering student, drives an old car (classifiable as an antique, since it is twenty-six years old). The odometer broke on at least one of the three or more previous owners, so the mileage is not certain, but judging from the wear on the throttle valve shaft, John estimates that it may be well on the way to a quarter of a million miles. Because the car has the stock electronic fuel injection, which has always run slightly lean, the carbon monoxide (CO) has always come in under 1 percent and the hydrocarbons are about 60 parts per million (ppm) when tested in New York state. The emissions are tested only at idle, on a warm engine. The compression ratio was originally 10.5:1 and is still around 8.8:1, not as high as it should be, but sufficient to provide OK performance. The loss of compression is caused by worn piston rings, and, as a result, the car burns enough oil under acceleration to stink, occasionally prompting an expletive from pedestrians and other motorists. Because motor oil is only a dollar a quart and he adds about that much only once for every five days of driving, John sees little reason to retire the car.

Questions for Discussion:

1. What are the facts in this case?

2. John's car may not be unique. Many older cars were overengineered and tend to die slowly, but the engines usually can be rebuilt easily. What should John do?

3. A researcher measured actual (real-time) emissions using an infrared beam, mirrored across a one-way road, and found that a nearly equal percentage of old and new cars were in the top 25 percent in CO and hydrocarbon emissions. Does this indicate inadequate emissions testing? Should the standards be revised? If new cars are "getting away with it," should John be able to pollute, too?

4. There are additives for older cars that, when put into the motor oil, are supposed to increase compression and decrease "blow-by." These solutions usually provide little improvement and are only temporary. There is also a risk of loosening gunk from elsewhere in the engine that might clog some of the smaller oil passageways and cause engine failure. Should John use one of these products?

> **1.2 Case Title:** It's in the Air (as)
> **Case Type:** Air Pollution

Fact Pattern

When Jim hears about the Montreal Protocol calling for the elimination of certain refrigerants damaging the atmosphere, he's worried. "I've spent my whole life working on refrigeration systems. I know these refrigerants inside and out. I don't think they're polluting the environment. What am I going to do if they eliminate chlorofluorocarbons?"

"Jim, get a grip," says Bill, his partner in J & B Automotive. "We'll just have to get prepared for the future. Let's get one of the new refrigerant manufacturers to come in and talk with us."

"You're right, Bill. Maybe I am overreacting. Perhaps this will be the best thing for us and our business," Jim replies.

When the HCFC salesman arrives, he extolls the virtues of the new refrigerants. "They're user-friendly. They're less toxic." The salesman continues, "Did you know that when you introduce this refrigerant into older cars, you'll have to perform a major overhaul, since the current O-ring seal material is not compatible with the new refrigerants?"

"Really?" says Bill. "Wow, I can see us making a lot of money on this."

"Yeah," says the salesman. "And, as the supply of CFC refrigerant starts to dwindle, everyone will have to switch to the new refrigerants because the cost of the older refrigerants will skyrocket."

Following the sales pitch, each of the mechanics is convinced. Jim rushes out to the local automotive supply house and buys twenty canisters of the CFC-based refrigerant. When he returns, Bill is furious.

"What are you doing?"

"I don't want any of my customers having to pay exorbitant prices for refrigerant," answers Jim, "and besides, it's better that this stuff is in the hands of responsible people like us, rather than in the hands of some do-it-yourselfer. They would probably release half the stuff into the atmosphere before they charged their AC system."

They both chuckled. "You're right about that," says Bill, "but it just seems like we're doing something illegal. And we're not going to make any money on this stuff now."

Questions for Discussion:

1. What are the facts in this case?

2. What are the choices available to J & B Automotive concerning the use of refrigerants?

3. What are the risks associated with each of the choices?

4. What are the benefits to J & B Automotive with each of the choices?

5. What are the risks and benefits to the customers of J & B Automotive?

6. Are each of the choices ethically equivalent?

1.3   Case Title: Selling Out the Environment (rs)
Case Type: Air Pollution

Fact Pattern

As an environmental engineering student, Thomas involves himself heavily in the affairs of the EPA, e.g., commenting on proposed regulations, attending EPA-sponsored seminars, et cetera. Thomas is concerned about two new proposals that have been brought to the EPA's attention.

Efficiency-Based Emission Factors
Cogeneration involves the simultaneous production of electricity and heat, in the form of hot water and/or steam. It is becoming more popular with utilities and independent power producers. Since this process recovers more of the energy from fuel as compared with conventional utilities, some parties are requesting that the EPA allow emission factors (AP-42 factors) to take into account the efficiency of cogeneration sites: If a site can generate more energy from the same amount of fuel, this fact should have an impact in determining the allowable emissions.

The Sale of Emission "Credits"
Advances in technology, and the addition of air pollution control equipment, can significantly reduce the amount of pollutants emitted from a facility. A new proposal to the EPA and state regulatory agencies includes allowing the sale of emission credits. (Some state agencies have already initiated regulations that allow for emission credits.) These emission credits will allow a facility that emits fewer pollutants than stated on their permit to sell emission credits to another facility who may be out of

compliance with their operating permit or to hold these credits for future use.

Thomas understands the magnitude of financial impact in the area of environmental compliance. However, he is undecided on the ethics behind these two proposals. Since he is more concerned about the environment than financial gain for businesses, Thomas proposes the following: If increased technology is available, those companies installing this technology should be required to reduce the amount of pollutants emitted from their facility, instead of taking financial advantage of the environment.

Questions for Discussion:

1.   What are the facts in this case?

2.   What are some of the problems Thomas may have with each proposal?

3.   What, if any, are the problems you may have with the proposals?

4.   What are your thoughts on the following question: Should benefits realized from increases in technology allow for facilities to install less pollution control equipment?

5.   What are your thoughts on the following question: Should facilities that install more efficient equipment or more pollution control equipment be allowed to sell emission credits to other facilities, which would in effect net out the environmental benefit?

6.   What are your thoughts on Thomas's proposal?

7.   What are your thoughts on financial impacts/gains from increased technology vs. environmental conservation?

2. *Solid Waste Pollution*

> **2.1 Case Title:** No One Else Does,
> Why Should I? (cp)
> **Case Type:** Solid Waste Pollution

Fact Pattern

Bob has been hired recently by a cleaning company whose contract calls for the nightly housekeeping chores in a large office building for a major international corporation. Bob is assigned to dust, vacuum, and empty the garbage in each of the offices and the common areas of two floors of this building. After about a week and a half, Bob has a pretty good routine down: how far he can reach with the extension cord to vacuum each section of carpet, how long it takes to dust each office, and where he can leave the garbage for systematic collection. He notices the amount of garbage he hauls to the basement Dumpster is quite large.

"This place certainly produces quite a bit of garbage," he thinks to himself, "and I'm only doing two of the twelve floors here."

Bob gazes over the almost-filled Dumpster, taking notice of the various things people throw out after a day's work in the office. Bags filled with boxes, magazines, paper, cans, bottles, and books are all mixed together. Bob wonders at what point this stuff gets separated in order to follow the recycling rules for the city. To satisfy his curiosity, after Bob has finished for the night, he approaches his supervisor, Jon, about the matter.

"Separate the garbage? Recycling laws?" Jon exclaims. "We don't do that here. It's not that I don't want to, but it's too much of a hassle to go and break down cardboard boxes, sort the junk mail from the good paper, and everything else."

"Aren't you afraid of what may happen if we're caught *not* doing it?" Bob asks.

"Look, kid, first of all, laws are made to be broken, right? Besides, I know of quite a few other cleaning crews in the city that don't separate their trash either, and I'm sure there are a lot more that also don't follow the city's rules. I have heard of only a handful of cleaning crews getting fined by the city for not separating the garbage."

"It doesn't worry you that there is the possibility of getting fined?" Bob continues.

"That's the beauty of it. Even if they do catch you and fine you, the fine is insignificant compared to what it would cost to hire more people in order to follow the city's rules. So you'll pay a few thousand dollars in fines. It's a lot better than paying $20,000 a year plus benefits to a new employee."

Bob is astonished at his supervisor's attitude. He can't imagine that there can be many other supervisors with Jon's attitude in this city. How could anyone *not* recycle after seeing the commercial of the crying Native American on television? Bob finds himself so caught up in the matter, he misses his stop on the train and has to walk five blocks to get home. During his walk, various ideas pop into his head about what can be done to better enforce the city's recycling laws: He could bring the matter to the attention of a police officer perhaps, write a letter to the mayor, send an E-mail to his congressman, or maybe organize a group to help enforce the rules. Then it dawns on Bob the amount of time and energy it will take on his part to do any of these things. What could one person possibly do? Bob begins to think, is it really worth it for him to worry about such things, even though he has learned all his life that recycling is something that *should* be done?

Questions for Discussion:

1. What are the facts in this case?

2. What is the issue in question?

3. Is Jon the supervisor correct in his attitude toward the law?

4. Is Bob correct in his feelings?

5. What do you think Jon may finally do?

6. What would you do if you were in Bob's place?

> **2.2**   **Case Title:** My Own Private Landfill (as)
> **Case Type:** Solid Waste Pollution

Fact Pattern

Alison has been working in the solid waste section of the Department of Environmental Conservation for four years when she lands an ideal job with the county. She has always wanted to live in the country, and the position of lead engineer in charge of the county landfill expansion will provide her with new opportunities and challenges.

"We're all sorry to see you go," says Tom, her supervisor at the DEC. "You've provided this department with a lot of great ideas, and your ability to get the job done is second to none. Having just received your professional engineering license is sure to be a big boost to your career."

"Thanks, Tom," answers Alison. "It's been a great experience here. I've learned so much about landfills that I'm sure there won't be anything in my new job that I can't handle."

"Have you found a place to live yet?" asks Wilma, her best friend at the DEC. Together, they discovered problems at the Quarry Rock landfill, where leachate from the landfill contaminated a nearby stream.

"Not yet, but I've been looking at the classifieds. I have a couple of places to look at tomorrow." Tomorrow is Saturday; her new position starts on Monday. Alison continues, "I know it's going to be rushed, and with all the Johnson State college students looking for apartments, it might be hard finding something I like that's affordable."

On Saturday, Alison takes the two-hour drive upstate, purchases the local paper, and looks through the classifieds again for any new listings. Then she calls a few of the promising locations. The first place she visits is nice, but they don't accept pets, and there is no way she's giving up her dog, Chipper. The second place looks nice on the outside, but when she goes inside, the apartment is trashed. She's about to give up hope when she stumbles upon an ad for guest quarters in an 1860 colonial. She calls the property owner and finds out that the quarters have not yet been rented, then quickly drives over.

When she arrives at the property, she takes an unpaved road, which seems to go on for miles, to the main house. The palatial estate consists of the main house and several smaller houses, which the owner tells her used to be occupied by servants when the house was owned by the town's mayor during the late 1800s. The owner, Mr. Teller, takes her to the guest house that is for rent. It's rather small, but the babbling brook adjacent to the house is quite soothing, and Mr. Teller has no problems with the dog—it will keep his twelve-year-old collie, Lad, company.

"I'll take it," says Alison. "You said all the utilities are included, right?"

"Yes, everything is included in the rental price," replies Mr. Teller. "Why don't you join me and the Mrs. for dinner tonight, after you get settled in? I can tell you about my career with the county."

"Thanks, but I think it will take me a lot of time to unpack what I've brought in my car, and I want to rest tonight. The movers are coming with my other belongings tomorrow. Perhaps we can have dinner another time."

On Sunday afternoon, after the movers have unloaded all her things, she starts to get settled. Mr. Teller knocks on the door and asks if she has any garbage to be thrown out.

"Well, actually, I have a lot of boxes, and some other junk."

"I'll take them for you," says Mr. Teller, who then disappears with the trash she has given him. Alison figures that the garbage pickup must be on Monday, so she quickly gathers the rest of her garbage and tries to catch her landlord. But she can't see him anywhere; not knowing where the garbage pickup is, she brings the garbage back to her apartment and leaves it outside.

Her first day at work is great. She's well rested, and she gains immediate respect at the landfill expansion with her broad knowledge of what seems like everything from surveying, to excavation, to hazardous materials.

When she arrives home from work, she notices that the garbage she has left outside is gone. She thinks to herself, "Well, garbage pickup is Monday, and I guess the trash collection point is right here." That evening, Mrs. Teller stops by with a garbage can and tells Alison, "Just put all your trash in here and leave it outside; we'll take care of it."

"She seems like a nice old lady," thinks Alison, and they start talking. Mrs. Teller reminds her of her own mother, quiet and reserved, yet smart and enchanting.

The next night, when she's coming home, she sees Mr. Teller's truck going down a side road on the estate with what looks like the garbage can that Mrs. Teller brought over the day before; when Alison arrives at the guest house, the garbage can is indeed missing. She gets back in her car and reverses her route, this time taking the side road that she saw Mr. Teller turn into. She quickly gets out when she sees what's happening.

"What are you doing?" screams Alison.

"I'm throwing out the garbage," says Mr. Teller.

"You can't just throw the garbage into a ditch, Mr. Teller." Alison moves a little closer to the edge and sees piles of garbage, including paint cans, household garbage, used oil drums, and other miscellaneous junk. She continues, "It's not right. You could be polluting that very stream that goes by the house."

"Listen, young lady. I've been putting my garbage here for the last twenty-five years. It's my property. I can do what I like with it. I don't need you to tell me what I can and cannot do on my own property."

Alison knows that what Mr. Teller is doing is wrong. But she loves her apartment, and after seeing how bad other rentals in the area are, and how few accept pets, she doesn't know what to do. She has already signed a lease for twelve months and paid two months' security deposit.

Questions for Discussion:

1. What are the facts in this case?

2. What is the problem that Alison is facing?

3. What are the risks that Alison might encounter if she reveals the Teller landfill?

4. What are the risks that Alison might encounter if she *does not* reveal the Teller landfill?

5. What are the risks to Mr. Teller if it is determined that the stream is contaminated?

6. What are the risks to *Mrs.* Teller if it is determined that the stream is contaminated?

2.3   Case Title: Packaging, Pollution, and Profit (zt)
Case Type: Solid Waste Pollution

Fact Pattern

SW Company, a chemical firm, has developed an inexpensive chemical specialty item, and Friedman, the manager of the company, hopes it will find a huge market as a household product. Friedman wants to package this product in one-gallon and two-gallon sizes.

A number of container materials would appear to be practical—glass, aluminum, treated paper, steel, and various types of plastic. A young engineer whom Friedman hired recently and assigned to the packaging department has done a container-disposal study that shows that the disposal cost for one-gallon containers can vary by a factor of three depending on the weight of the container, whether it can be recycled, whether it is easy to incinerate, whether it has good landfill characteristics, et cetera.

SW Company's marketing expert believes that the container material with the highest consumer appeal is the one that happens to present the biggest disposal problem and cost to communities. He estimates that the sales potential would be at least 10 percent less if the easiest-to-dispose-of, salvageable container were used because this container would be somewhat less distinctive and attractive.

Assuming that the actual costs of the containers are about the same, Friedman must make a decision.

Questions for Discussion:

1.   What are the facts in this case?

2.   What is the ethical dilemma?

3.   What would you do if you were in Friedman's situation?

4.   What will Friedman do, in your opinion?

3. *Water Pollution*

> 3.1   **Case Title:** Lead in the Pool (sc)
> **Case Type:** Water Pollution

Fact Pattern

Eddy is an engineer working for a small environmental consulting firm. His firm has been hired by a local swimming resort to analyze the effluent water from old pipes in order to determine the amount of organics present in the water. The resort owner hopes to eliminate the organics by treating them at optimal points along the water's path. Eddy's task is to find the various concentrations of organics along the pipelines.

Knowing lead to be a very toxic metal and corrosion of old lead piping to be a primary source, Eddy also performs tests to analyze the lead concentration. The results of the tests show that the lead in the pipe is more than double the proposed EPA concentration limit. From experience, Eddy knows that the only way to eliminate the lead is to replace the pipe.

Before writing up his report, Eddy informs the owner of the resort of his findings: "I analyzed the lead content in your pipes and found that the levels are higher than the EPA regulations."

"I didn't hire you for that. I know that there's a lot of lead in those pipes, but if you think I'm replacing them, you're crazy. Do you know how much that would cost me?" replies the owner.

Eddy realizes he shouldn't have taken it upon himself to analyze the resort's pipes for lead, but he did, and now he doesn't know what to do. Lead is toxic. It has been proven to cause all kinds of biological damage. In any event, the owner isn't going to get rid of the problem, and Eddy wouldn't have known about the problem if he'd only done what he was hired to do.

Questions for Discussion:

1. What are the facts in this case?

2. Should Eddy have analyzed the pipes for lead?

3. Is the owner's response to Eddy's findings logical? ethical?

4. Based on his new knowledge, is Eddy responsible for the people who use the water at the resort?

5. What responsibility does Eddy have to his employer?

6. What responsibility does Eddy have to the reputation of his firm?

3.2   Case Title: Killer Catches (bd)
Case Type: Water Pollution

Fact Pattern

A hotel owner in a developing country, depending on tourism for most of his business, is seeking an economical means of providing pure water for all guest consumption. There is an ample supply of water available in the town through a reasonably reliable pipe system, and at very low rates. That system is now being used for all sanitary purposes except drinking.

Sam, a visiting civil engineer for the company hired to conduct the project, has proposed to the owner that he install an automatic filtration and chlorination plant and process all the water he uses. After a more careful investigation, however, Sam uncovers a devastating fact. He learns that local fishermen have been dumping deadly sodium cyanide into the waters to stun large fish so they can be easily captured for the restaurant and aquarium trades. The toxin often kills small fish and corals, transforming shallow reef communities into aquatic graveyards. Sam immediately informs his supervisor, John, of his unexpected findings and asks if he should report this information to the proper authorities.

John advises Sam not to mention this information in his report, but to perform only the services for which the company has been hired.

"But John, hundreds of reefs have been affected, and there's no end in sight," says Sam. "Seriously damaged reef communities sometimes take several decades to recover, even under the most favorable conditions."

"I understand your concerns, Sam, but it is unlikely that these violated reefs will ever again be allowed to

flourish," says John. "Poor villagers in the region probably count on reef fish to feed their families and earn a little extra spending money, too."

"Maybe you're right," Sam utters.

"I know I am right," states John. "Besides, disclosing
this information is not good for business and under-
mines principles long established in successful business
operations."

Questions for Discussion:

1. What are the facts in this case?

2. Should Sam report his findings to the proper persons?

3. Do you agree with John's advice?

4. What do you think will be the final outcome?

3.3   Case Title: An Ocean of Trouble (jmsl)
Case Type: Water Pollution

Fact Pattern

The study of marine habitat has always interested and excited Bill. He is very happy in this profession and delights in the fact that no two days are ever alike.

Julie is an engineer whose work combines marine studies with mechanical engineering.

Today they are going out into the ocean to study a new form of power generation. The project is based on their belief that it was possible to harness the energy present in waves to generate electricity. The team includes four research engineers and two boats. The plan is to get out to the site by 9:00 A.M. so they will have enough time to accomplish all of their work and return to port a little after dark.

As it turns out, they are ahead of schedule and make it to the site at quarter to nine. The first thing they do is set up the equipment.

"I really wish we could have gotten a nicer day for all this work we have to do," states Bill.

"Yes, it sounds ironic, but too much wave height can really interfere with the work we have to do today," replies Julie.

"The weather report said that it's only going to get worse. The seas will be building to almost six feet by the time we leave," remarks Bill.

Once the equipment is in the water, the work proceeds rather smoothly. They encounter no major problems and gather a significant amount of data that will be extremely useful to the project when it is analyzed back at the university. Bill, the team leader, decides when it is just about time to conclude work for the day and remove the equipment from the water. Unfortunately this must be done rather quickly because the weather is deteriorating at a rapid rate.

Just as they are about to remove the last piece of equipment, something goes wrong. One of the hydraulic lines from the experimental apparatus gets snagged on one of the deck cleats. In an instant there is a large volume of hydraulic oil sprayed out all over the place. Luckily, the apparatus is not yet inside the boat, and the oil does not cover all of the boat's expensive instrumentation.

"We've gotta get the oil flow stopped!" exclaims Bill.

"It's going all over the place. I never knew that this thing contained so much hydraulic fluid," interjects Julie.

"Our hands are tied; the piece of equipment is still on the hoist, and the only way to stop it would be to bring it on board and manually turn off the supply of hydraulic oil. The weather isn't helping us, either. We'll just have to wait until the oil runs out," states Bill.

"There's so much oil in there. It could take hours. What about the tremendous oil slick that we'll be creating?" asks Julie.

"Like I said, there's nothing we can do about it," replies Bill.

"Well, at least I think we could notify the Coast Guard about the oil spill. They can send someone out here to contain the spill and prevent it from doing any harm," states Julie.

"Under most circumstances I would agree with you, but do you know what effect this could have on our project? The cleanup costs alone could easily use up all of our funding. It's dark right now, and there will be no way to tell that it was us who spilled the oil. I think it would be in our best interests to just leave," answers Bill.

Questions for Discussion:

1.   What are the facts in this case?

2.   Should the project have proceeded even though Bill knew that the weather conditions were going to deteriorate?

3.   What problems face Julie at the end?

4.   Is Bill's reasoning for not calling the Coast Guard correct?

5.   What effects may this incident have on future research efforts?

D. ETHICAL CONFLICTS INVOLVING FRIENDS, FAMILY, AND COWORKERS

1. *Air Pollution*

> **1.1** **Case Title:** Freeing Freon Is
> Costly Business (kjc)
> **Case Type:** Air Pollution

Fact Pattern

Donald, an auto mechanic, has worked for Engle's service station for five years. He often has to repair or recharge air conditioning units on older cars. Often, especially during the summer months, Donald has to repair fifteen or twenty units a day. With this kind of volume, the Freon recycling unit gets a lot of use. Due to the recovery procedures and the amount of use it gets, it is very susceptible to damage.

After a hard day at work, Donald gets together with some friends to relax; in doing so, he begins to tell of how the recycling machine is broken often, causing him and the station to lose business and money to their neighborhood competitor.

Greg is one of Donald's childhood buddies and an environmental engineer for the EPA. He knows that Donald has been instructed to leak Freon into the atmosphere, and often does, so that he can service more cars and steal business away from the garage down the street. Since this is cheaper than the proper procedure, both the garage and Donald make more money.

Greg is faced with a difficult choice.

Questions for Discussion:

1.   What are the facts in this case?

2.   What are the ethical issues in this case?

3.   What is the problem Greg is faced with?

4.   Is Greg putting his own job in jeopardy if he keeps quiet?

5.   How, if at all, does Greg's decision change if Donald is struggling to support his wife and newborn child?

1.2 Case Title: A Breath of Fresh Air (jg)
Case Type: Air Pollution

Fact Pattern

Sara is a junior engineer for Fresh Air Inc. After being with the company a little over a year, she has finally gotten her "big break." She is now a key player in a bid for a contract from Stack Emissions Inc. She has put in tons of time and work getting together the proper information as well as making contacts. If she lands the contract, there will definitely be a sizeable raise and promotion to senior engineer. Sara is hungry for the promotion, not to mention the money. Stack Emissions Inc. raved about her efforts until recently, when another company stepped in and created a bidding war. This is frustrating to Sara because of all the hard work she has done. She is also getting tons of pressure from her boss.

"Sara, if we lose this contract, the company will have to make some cutbacks," her boss grumbles. "We simply cannot afford another loss. I want you there selling this company. I have no doubts that your work has been exceptional, but something more must be done. We need to compete with the other bid, and you need to be on that now!" he demands as he leaves in a huff.

Sara is at a loss for options on how to proceed. She decides to see a friend from college for lunch to clear her head and get a little advice. Jane is another engineer working for a similar company. When Sara arrives, Jane is not at her desk, so Sara wanders around the office looking for her. She's familiar with most of Jane's coworkers, and she greets them pleasantly as she is looking. As Sara walks toward the copy room, she sees a Stack Emissions letterhead on a piece of stationery with files scattered beneath it.

"Could Jane's company be the other bidder?" she wonders. "This project is right up their alley. I remember when Jane first started, they had her working on something very similar."

Now Sara's mind is racing. She finds herself staring at the pile on the desk trying to make out as much as possible. And there it is, the project number!

"That's it! They are the other bidder. I've got to find out what their estimates are. If I can get those numbers, then I can easily beat their price and get the contract." Sara is now looking around her to see who is where.

She sits in the chair at the desk, which obviously belongs to a secretary, who is probably out to lunch herself. She picks up the phone and dials her machine so that she looks as if she has a right to be there. As she listens to her voice on the other end, she casually opens a file. In front of her is all the data she too has worked on compiling all these months. The work is very similar to hers, and she even spots a few errors. At the very bottom she sees a file marked BUDGET. Bingo!

Questions for Discussion:

1. What are the facts in this case?

2. What is the dilemma with the contract?

3. What is in jeopardy if the contract is lost?

4. What is Sara trying to find at the desk?

5. What would you do if you were put in Sara's position?

6. What are the implications if she gets caught?

7. If Sara gets the budget numbers and is able to win the contract from peeking at the file, do you feel it is the right thing to do?

8. Explain your decision for question 7.

1.3   Case Title: Midnight Oil (jmsl)
Case Type: Air Pollution

Fact Pattern

Mark runs a successful heating oil company, established for over thirty years and handed down by his father. Essentially, all of their business is generated through the selling of fuel oil for heating purposes and through the service of their customers' oil burners.

Jim, Mark's brother, has chosen not to get involved in his father's business, but instead starts a business of his own. His interest in automobiles sparks his desire to open up a service station, which becomes quite successful, just as Mark's business has been.

Recently, however, Jim has faced some problems. There have been increasing concerns about used petroleum products. The local, state, and federal governments have mandated strict instructions on how to handle such things as used motor oil and unusable gasoline. Unfortunately, as Jim discovers, the costs of compliance can sometimes tax a small business, even drive it to the verge of financial instability. Jim decides to pay a visit to Mark.

"Hey, Jim, how's everything? I haven't heard from you in a while. Is everything all right?" asks Mark.

"Oh sure, everything is great, things couldn't be better," replies Jim.

"Well, things are good here. As usual, business picks up in the fall, like now, and goes well straight through the winter," states Mark.

"I guess things are pretty good with me too. We do hit some slow spots, though. It seems that they last forever, but I guess that'll end soon," says Jim.

"Yeah, I know. It does happen occasionally. If there's anything I can do to help you through it, just let me know. I'd do anything to help you, and I know that you'd do the same for me," states Mark.

"You know, things would be so much better if I didn't have the EPA breathing down my neck about all of these so-called toxic substances. It seems like you can't do anything or use anything without having to pay someone to dispose of the 'toxic' substances created in the process," explains Jim.

"Well, I guess I'm lucky in that respect. The EPA knows we're clean and leaves us alone because of that," replies Mark.

Jim is leading into the topic that is his true reason for coming here. He once heard of an incident where waste petroleum products were added to clean heating oil in order to dispose of them. This would allow Jim to get rid of the "nasty stuff" without having to pay someone to take it away. There would be no possible way that anyone would find out what they're doing. It's time to ask his brother for help.

"You know, now that I think of it, there is something that maybe you can help me with. As a result of our daily operations, we generate a significant amount of waste oil, like when we do oil changes. I have to pay someone to truck this old oil away. The EPA says that I can't charge people to accept their used oil, so I lose money. It's just sucking me dry," states Jim.

"I don't see where I can help," replies Mark.

"Well, I once heard that people were taking used petroleum products and adding them to fuel oil in order to dispose of the oil. It wouldn't cost a dime, and your customers won't even notice," explains Jim.

"I don't know if that is such a good idea; there's too much at stake," says Mark.

"Come on. When they take away my old oil, they just incinerate it anyway. We would be doing the same thing. It's harmless, I promise," states Jim.

Questions for Discussion:

1.   What are the facts in this case?

2.   What environmental issues face Jim?

3.   What are the reasons for Jim not wanting to dispose of the oil properly?

4.   What are the possible implications of adding unknown petroleum products to oil used for heating?

5.   How does Jim and Mark's family relationship affect this situation?

6.   What are some of the consequences that face Mark as a result of his impending decision?

2. Solid Waste Pollution

2.1   Case Title: Your Job *or* Our Friendship (ec)
Case Type: Solid Waste Pollution

Fact Pattern

John and David have been best friends since college, where they both studied environmental engineering. After graduation, John starts his own, relatively small R&D firm, FutureTech. David proceeded on to study environmental law. Today, they both work in the New York area and remain close.

Thursday, David is at work, and his boss calls him into his office. David receives a new case that involves cracking down on a company illegally dumping solid waste into nearby waters. To David's dismay, he discovers the company is FutureTech. Despite his mixed feelings about the case, he knows it is his job to continue with the case.

The following week, David goes to FutureTech and meets with John, telling him about the case and how he is in charge of it.

"I cannot believe this," proclaims John. "After all our years of friendship, you are going to sue my company?"

"This has nothing to do with our friendship," responds David. "Your company is involved with illegal dumpings, and it is my job to see to it that it stops."

"I could lose everything. I promise it will stop. Tell your boss that it actually is not my company involved in these illegal actions. I swear to you, if you go through with this, I will no longer consider you my friend."

"I cannot believe you are putting me in this situation," exclaims David. "I am only doing my job."

"Well it's your job *or* our friendship," answers John.

Questions for Discussion:

1.   What are the facts in this case?

2.   What are the complicating circumstances in this case?

3.   Are John's actions ethical?

4.   What do you think David will do?

5.   What should David do?

2.2   Case Title: Walking on Broken Glass (rc)
Case Type: Solid Waste Pollution

Fact Pattern

A material recovery facility located in Anytown, New York, has been in operation for ten years, and has an outstanding record for performance. In all of the years of its existence, the MRF has offered a product that can be used by other industries at relatively high efficiency. The present facilities at the plant allow for the processing and separation of paper, cardboard, plastics, and glass. Under these operating conditions, 85 percent of the materials (by weight) entering the facility are sold to companies that will make use of the recycled materials. The remaining 15 percent consists of unrecyclable plastics and glass that has been broken; as a result, the pieces are too small to be sorted by the current technology employed at the plant. Clear, green, and brown glass must all be separated to be reuseable. If some type of new technology could be employed at the facility to separate the broken glass according to color, an additional 10 percent of the materials (by weight) could be recycled.

Mr. Robert Smith is in charge of research and development at the Anytown MRF. The success of the MRF can largely be attributed to his hard work and knowledge of how a MRF is to be operated. In an attempt to make the MRF at Anytown a truly state-of-the-art facility and also bring in more revenue, Mr. Smith has been looking into obtaining a grant from the state to begin testing a new piece of machinery—an optical sorter—that would allow the plant to sort even the smallest pieces of glass. The optical sorter employs a laser "eye" that determines the color of the glass and then sorts the pieces accordingly.

Almost all of the MRFs in the state will be bidding to obtain this grant to employ such a sorter at their facility.

In order to increase his chances of obtaining this state grant, Mr. Smith has a plan. He decides to contact Mr. Tom Jolly, a longtime friend of his at Glass Company, to whom the MRF at Anytown sells recovered glass. Mr. Smith explains to Mr. Jolly that if he could use his resources to pressure the state planning board into giving the research grant to the MRF at Anytown, he would ensure that Glass Company would be locked into a low price for recovered glass.

"Tom, I am asking you only to put a good word in with the state board. Let them know about who we are and what type of operation we run. You have contacts all over the state. I'm sure you could easily make it known to all of them that it would be in everyone's best interests for our facility to receive the grant. In return, your company would be fixed at a low price for recovered glass, regardless of the current market value."

"I am honored that you believe that I have such a large influence over the current affairs in the state," explains Mr. Jolly. "I can let them know about the type of operation that you have put together, but I cannot make a recommendation of your particular plant, nor can I accept any kind of favorable status."

"It's not favoritism, it's intelligent and aggressive networking strategy," responds Mr. Smith.

Questions for Discussion:

1. What are the facts in this case?

2. What are the moral implications of Mr. Smith approaching Mr. Jolly with his plan?

3. Is Mr. Smith jeopardizing his friendship with Mr. Jolly?

4. Legally, is there anything wrong with Mr. Smith's plan?

2.3 Case Title: No Dumping Allowed (robp)
Case Type: Solid Waste Pollution

Fact Pattern

Mike is a technician in a wastewater treatment facility. He has recently been put in charge of disposing all of the sludge that is generated by the facility. Although he is not familiar with the legal aspects of this process, Mike calls both state and federal agencies and discovers that he has two options. He can either sell the sludge to farmers for use as fertilizer or dispose of it in landfills. He is also informed that the dumping of sludge into the ocean is no longer legal.

After weeks of sludge disposal, Mike has discovered that it is very hard to sell sludge on the open market, since most farmers already have contracts with wastewater facilities to buy their sludge. Mike has also discovered that landfills charge on a per ton basis for the dumping of sludge; due to the large amount of sludge produced by the facility, it is very costly.

Although Mike has not been on the job long, he realizes that if he continues to spend money at the same rate for the disposal of sludge, he will grossly surpass the facilities budget. Mike, not knowing how to deal with this dilemma, decides to contact the previous technician, Joe, who was in charge of the disposal of sludge.

"Well, Mike, I can solve your problem for you. All you have to do is dump your sludge into the ocean."

"Into the ocean? That's illegal. They stopped doing that about five years ago," replies Mike.

"Some of the facilities stopped. However, we weren't one of them," states Joe. "That law claims that the dumping of sludge affects the biota of the ocean, and that's not true. Sludge has been dumped into the ocean for years."

"Does the boss know about this?" asks Mike.

"Of course not. And don't tell him. If he knew, I might lose my pension," replies Joe. "I'll give you the number to call, and the people there will take care of you."

"I don't know if I should do that," states Mike.

"If you don't dump the sludge my way, you will spend much more than your allotted budget, and that will not look good for you," Joe comments.

Questions for Discussion:

1. What are the facts in this case?

2. What problem is Mike facing?

3. If Mike explains the situation to his boss, what reactions do you think the boss will have?

4. What about the previous technician's comments? What type of mind frame does he have?

5. If you were faced with the dilemma Mike has, what would you do?

3. Water Pollution

> **3.1   Case Title:** My Lips Are Sealed (yf)
> **Case Type:** Water Pollution

Fact Pattern

Billy is an engineer at Chic Environmental, Inc., whose supervisor assigns him to a project dealing with air strippers in a nearby locality. The air strippers have been added to the local water treatment facility, LaLa wastewater treatment plant, as a supplemental purification process. The exiting gas stream from the stripper meets all required permit and legal regulations, so the only issue that Billy must address is increasing the efficiency of the system without causing a major aesthetic upset.

In order to gather data on the composition of the water that comes into the air stripping towers, Billy asks his buddy Jimmy, who works at WaWa wastewater treatment plant, just upstream of LaLa, if he can supply the necessary reports and data for Billy's project.

"I am going to need your monthly flow reports, including the maximum and minimum flows, in order to develop a model of the stream as it proceeds down from the WaWa plant to the LaLa plant. By developing a model, I will be able to accurately determine the makeup of the stream and then the best method for increasing efficiency once it reaches the air stripping towers."

"I can give you all the information you need," says Jimmy, "but I feel I should let you in on a little bit of classified information before you go crazy trying to do this project. His voice drops to a whisper. "To be honest with you, Billy, I know for a fact that the Bando Water Company, which is located upstream of the WaWa plant, draws from this stream during times of high flow. Every once in a while, during low flow, you will hear the Bando pumps buzzing, which means that the company is drawing water when it shouldn't. This definitely has an impact on the stream flow and composition. As you already know, low flow equals limited dilution!"

Bill replies, "If this ever got out, Jimmy, the public would be outraged! I'm sure the water is cleaned before being used, but you know how scared the public gets about infection and cleanliness. I think they have the right to know by what means their water supply is being obtained. The media would be certain to pick up on this, and it would mean a lot of bad publicity. An investigation of all potentially informed parties could occur, and both our companies could be under close watch for quite some time! And . . ."

"Calm down!" interrupts Jimmy. "No one is going to know anything if you keep this to yourself. Just remember, you didn't see it, so you don't know about it. I just thought you should be aware of it before you start this project. I'm giving you this information in confidence, so please keep it that way."

After a small investigation of his own, Billy finds out that the LaLa plant has been having trouble meeting permit requirements on certain days of low flow. Coincidentally, it seems to occur on the same days that the Bando Water Company draws its water. He begins to wonder to himself if he should mention anything to the personnel at the LaLa plant. If he does say something, he will surely be praised by both the plant and his company. However, it would rankle his conscience that he broke his word to his pal Jimmy. Alternatively, if he does not mention anything, he would have to fudge some numbers in order to complete the project in an acceptable manner. However, by not telling, he would always know in the back of his mind that he is deceiving the public and is not doing his job as well as he could if he divulged this information.

Billy goes to a colleague for advice. "The right thing to do in this situation is to inform the LaLa plant about the Bando Water Company," says that fellow worker and trusted confidant. "You could keep the source anonymous, and this way keep your friendship with Jimmy, yet feel better about yourself."

Questions for Discussion:

1. What are the facts in this case?

2. Was it ethical for Jimmy to "rat" on the Bando Water Company?

3. Should Billy have gone to a colleague (thereby divulging the information to yet another party), or should he have gone straight to his supervisor?

4. Should Billy take the allegations into account as he performs the project?

5. Would it be ethical for Billy to go to the LaLa plant and reveal the information passed on to him?

6. Is it a betrayal of confidence to leak the information when people's health is at stake?

7. Should Jimmy be the one to speak up, since he is the witness to the wrongdoing?

> **3.2 Case Title:** Protecting Familial
> Ecology (bg)
> **Case Type:** Water Pollution

Fact Pattern

Randolph Bazile, who has worked for the Environmental Protection Agency (EPA) for seventeen years, is the team leader of thirty people. Their duty is to check the concentration of nutrients in the effluent (treated water that is then discharged) from various wastewater treatment plants. If the concentration of nutrients is above the standards of the EPA, that agency issues an order to the plant in question to change from secondary treatment to tertiary treatment of wastewater.

Randolph's son, Greg, follows in his father's footsteps and majors in environmental engineering. In several months, Greg will begin working at the Metro WasteWater Treatment Plant. Randolph's team recently tested the effluent of this plant for large concentrations of nutrients such as nitrogen. The EPA's standard is that nitrogen concentrations must not exceed the regulation limit because the resulting increase in phytoplankton leads to eutrophication, which produces poor water quality.

Randolph's team found that the level of nutrients was below the specification of the EPA. However, because Randolph's son will be working at this plant, he suggests to the plant manager that he begin projects on developing tertiary treatment. Although this would increase taxes in Metro, it would also increase the number of jobs for environmental engineers, increase job security for future employees, and increase their salaries as well.

Questions for Discussion:

1.   What are the facts in this case?

2.   Should the plant use taxpayers' money to develop tertiary treatment even though it meets the requirements of the EPA?

3.   Even though society will benefit from tertiary treatment, is it fair for Randolph to use his position to secure his son's job and increase his salary?

> **3.3   Case Title:** Just a Teardrop in the Ocean (bh)
> **Case Type:** Water Pollution

Fact Pattern

Anita is a chemical engineer who has just been hired out of college to work for ACME Electrical Company. It is her job to monitor the amounts of waste that her particular plant "legally" discharges into a local river system. ACME Electrical is in the business of manufacturing large capacitors and transformers for power stations. The company requires the use of polychlorinated biphenyls (PCBs) as the fire-retardant fluid in their capacitors; PCBs have been known to cause health problems, such as nervous-system disorders and cancer, in aquatic life and in some humans.

Since the plant is located near a large commercial fishing village, the PCB levels in its waste effluent must be watched closely. Several years ago, ACME Electrical was cited by the EPA for contaminating the river with as much as 1.1 million tons of PCBs during a thirty-year period. The Superfund Act required ACME to pay for the entire cleanup effort to rid the river of the contaminant, setting the company back financially. Needless to say, the company was not happy about this turn of events and made it clear to its employees that a repeat would mean "some heads would roll!" During the last few years, the company's waste to the river has consistently passed regulatory inspections, and the EPA has basically left the plant alone to conduct its own business.

Anita's first few months at her new job are fairly uneventful. During this time she never observes a PCB level in the plant's waste above 0.0005 mg/l, which is an acceptable maximum contaminant level (MCL) by EPA standards. However, Anita *does* begin to notice a slight increase in the PCB levels every week, but since these increases remain within the desired MCL, she doesn't mention it to her supervisor. One day, though, the levels shoot up to almost twice the allowable concentration for PCBs. Anita is well aware that, for not reporting the gradual increase in PCBs sooner, she could go from being the "new kid on the block" to the first kid on the "chopping block." Unsure of what to do, she decides to confide in Stan, one of her trusted coworkers.

"Well, for how long did you notice that the PCB levels were increasing?" asks Stan.

"Not for more than a month. But as I said, I didn't report the increase because the concentration levels were still acceptable," replies Anita. "What should I do? Technically, I did what I was supposed to do. How was I supposed to be able to predict that the PCB levels would double in just over a week?"

"Yeah, but technically that's still enough to get you fired in the company's eyes," Stan comments sarcastically. "I say that if no one has found out about it by now, then no one will. The EPA has forgotten all about us since the cleanup, and the company executives would rather know nothing about it. Besides, if it happened only once, it's a mere teardrop in the ocean. Believe me, you don't want to throw your career away for such an insignificant mishap."

"I suppose you're right, Stan," Anita says solemnly. "Thanks for the help."

Questions for Discussion:

1. What are the facts in this case?

2. What kind of problem is Anita facing?

3. What consequences could Anita and ACME Electrical Company face if she reports the increase in PCB concentration?

4. What consequences could Anita and ACME Electrical Company face if she *doesn't* report the increase in PCB concentration?

5. Will following Stan's advice help or harm Anita?

6. What final decision do you think Anita will make?

2

Domestic Applications

2.1   Case Title: Who's Gonna Know? (kjc)
Case Type: Domestic Applications

Fact Pattern

John has made a career decision and bought a store. He intends to open a pizza parlor in a prime location on the main strip in his neighborhood. The store that previously occupied the location was a dry cleaner.

While John is remodeling, he discovers large drums that were left from the previous business, some of which are labeled benzene. John inquires about the proper disposal procedure and finds that even if he can somehow justify the exorbitant expense of taking care of the former owner's problem, it will still take about four to six months of paperwork to get rid of these drums. Thus, he will lose time and money because he cannot finish his renovations and open for business. A friend suggests that since John already is discarding trash in a large refuse container while renovating, he should just put the drums in it and forget about them; after all, that is what the dry cleaner did. Nobody would even question what was being thrown into the Dumpster anyway.

John makes a big decision.

Questions for Discussion:

1.   What are the facts in this case?

2.   What is the problem John is faced with?

3.   Who is John putting in jeopardy if he just disposes of the drums quietly and illegally?

4.   What if John just sells the store without opening and without telling anyone about the containers?

5.   What do you predict John does?

> **2.2   Case Title:** Hazardous Leftovers (rtc)
> **Case Type:** Domestic Applications

Fact Pattern

Peter and Marie, a young couple who have two children and many existing demands on their budget, have just purchased their first house. The house is very old, certainly a handyman's fixer-upper; therefore, on top of their mortgage payments, a lot of money will be spent on repairing and remodeling. Faced with all of these bills, Peter and Marie are counting their pennies and making their best attempt to keep costs to a minimum.

The couple have bought this house from an elderly widow who did not do much for the upkeep, so the house and property collected a great deal of junk. Since it is summer, one of the first projects the couple decides to undertake is to clean the property so that their children might play outside in the nice weather. The yard is filled with rocks, weeds, old wood, and miscellaneous cast-off materials.

One of Peter's closest friends, Matt, is an environmental engineer who used to work in construction before graduating college. As it turns out, Matt is the first person Peter asks for help in clearing the property.

Peter's plan is to throw all of the junk in his truck and bring it to the local dumping station, which will then dump the garbage in the nearest landfill. As the two men are cleaning the yard, they come across some old, greasy wood—large railroad ties used for making retaining walls, immediately recognized by Matt.

"Hey, Peter, do you realize that these are treated with creosote?" Matt says.

"I don't have a clue what creosote is," replies Peter.

"Creosote is a chemical that was once used to treat wood to protect it from termites. After being used for a while, it was found to be very dangerous to the environment. We can't just throw this stuff into a landfill."

"What should we do with it?" asks Peter.

"I'm not too sure, but let's try calling the state Department of Environmental Protection to see what they say," Matt answers.

Peter responds, "No way. I'm not calling those environmental guys because it's going to cost me a ton of money to throw this stuff away. Besides, there isn't much wood here, it won't make much of a difference. The guy at the dump won't know, and nobody else will either."

Questions for Discussion:

1.   What are the facts in this case?

2.   What are the issues faced by the two men?

3.   To what or whom does Matt owe his loyalty—his career or his friend?

4.   What is the dilemma facing Matt?

5.   What is the dilemma facing Peter?

6.   Since Matt is doing Peter a favor, should Peter be understanding and let Matt make the decision as to what they should do with the wood?

2.3 **Case Title:** Faulty Asphalt (sc)
Case Type: Domestic Applications

Fact Pattern

Hilltop Construction has been contracted to rebuild a few main roads in the small town of Downtrodden. Hilltop is a large contracting company and, as such, is able to underbid each of the smaller companies to get the job.

Bob, the engineer in charge of the project, is approached by his supervisor, Frank, and is told that on this job a lower-quality asphalt than the Hilltop norm is to be used. Bob inquires about the reason for the change.

"Well, Bob, this is the first time this town has had work done on their roads since who knows when. They know nothing about asphalt quality," says Frank, "so we can give them anything. If we use top-grade asphalt, we'll have to pay more for materials."

"You mean like when we do jobs in the city?" asks Bob.

"Listen, Bob, this isn't a city job. These guys won't even know the difference; besides, the asphalt isn't going to be unsafe, it just won't be of the same quality that the more experienced clients demand."

Bob doesn't feel like arguing, so he just shrugs it off. He knows it's unfair, but life's unfair and what can he do about it anyway? It's not as if he is the one in charge.

Questions for Discussion:

1. What are the facts in this case?

2. Should the town of Downtrodden receive the same service as the more affluent and educated towns/clients?

3. Is Frank's reasoning justifiable?

4. Does Bob have any control over the situation?

5. Should the people of Downtrodden be any of Bob's concern?

> **2.4 Case Title:** Cheap at What Price? (pf)
> **Case Type:** Domestic Applications

Fact Pattern

Zurich is one of the gardeners on an estate. The head groundskeeper requests that Zurich fertilize the grounds. When Zurich opens the fertilizer, he notices that it is a much cheaper brand than the one normally used; he also notices that there is a warning on the bag about potential harm to animals and advice to keep small animals off lawns for two days after fertilizing them. Zurich asks the head groundskeeper if the fertilizer was bought accidentally, but he is told that the owner of the estate purchased it to save money. Zurich then asks the *owner* of the estate if he should return this fertilizer and purchase the usual brand because of the warning he has read on the bag of the new kind.

"No, you shouldn't return it. I purchased this fertilizer because it's cheaper," states owner Dan.

"But, what about the warning for the animals?" asks Zurich.

"What animals? I don't have any pets. Now quit bothering me and get back to work," barks Dan.

Zurich still has qualms about laying the fertilizer, fearing a neighbor's pet may get ill. Yet, since there is a chain-link fence surrounding the property, Zurich believes the fertilizer won't cause any harm. That night, though, there is a heavy storm, and the fertilizer is spread to the neighbor's property by the wind and rain. The next day the neighbor's child becomes ill after playing in the grass. When Zurich hears that news, he fears that the child has become ill from the fertilizer. He informs Dan of his concern.

Dan has already spoken with the child's parents, who were informed by their doctor that the child would be fine in a matter of days, although the doctor isn't exactly sure what has caused the illness. Dan, therefore, does not see the need to tell the neighbors that the fertilizer may have been the cause of the child's stomachache. In reality Dan is concerned that the neighbors might sue him if they believe the fertilizer is to blame. Dan thus warns Zurich against telling the neighbors about the possibility of the fertilizer having caused the harm.

"What's the sense in telling them about the fertilizer? The child is fine now, and no permanent harm was done. Who's to know if it was the fertilizer that caused the illness anyway? Therefore, Zurich, I don't want you to mention anything about the fertilizer to our neighbors," Dan states authoritatively.

Zurich knows that the child will regain full health, yet he feels it deceitful not to inform the neighbors of the possibility that the fertilizer is the cause of the child's illness. He knows that he might be in trouble with Dan if he tells the neighbors what happened. Yet Zurich feels guilty and decides he should tell the neighbors just in case the fertilizer might have some unknown side effect or cause long-term damage. He promptly informs the neighbors of the possibility that the fertilizer has caused the child's illness. The neighbors then receive word from their doctor that the fertilizer *did* cause the child's illness, and they decide to sue the owner of the estate. Upon learning that he is being sued, the owner of the estate proceeds to fire Zurich.

Questions for Discussion:

1. What are the facts in this case?

2.   What ethical dilemma has occurred?

3.   In what way is Dan, the owner of the estate, to blame for the child's illness?

4.   Is Zurich responsible for the child's illness in anyway?

5.   If Zurich had been able to determine that the fertilizer would have no side effects or do no permanent damage to the child, should he have told the neighbors?

6.   Is the owner of the estate justified in firing Zurich?

7.   Are there other actions Zurich could have taken at any time after his first contact with the new fertilizer that might have had better results for the child and himself?

```
┌─────────────────────────────────────────────────┐
│                                                   │
│   2.5   Case Title: You Wouldn't Do               │
│                     That to Your Father,          │
│                     Would You? (jm)               │
│         Case Type: Domestic Applications          │
│                                                   │
└─────────────────────────────────────────────────┘
```

Fact Pattern

It has been one of the hottest, driest summers in recent memory. The local reservoir is on the brink of being empty, and it is only midway through July. Steve is the geohydrologist in charge of all the calculations for the reservoir, such as the safe yield, design capacity, and so on. He is asked to analyze the near-empty reservoir and express his thoughts on what can be done to salvage the situation.

Steve begins his analysis: Every three days he calculates how much volume is lost due to domestic use and lack of precipitation. At the beginning of August, there still has been no rain, and the reservoir volume is now dangerously below its minimum capacity. Steve reports to his supervisor that the only way to remedy the situation is to mandate that residents cut down on domestic water use.

Steve decides that, for the time being, each person will be restricted to using a maximum of ten liters of water per day, or forty liters per day for a family. That would mean no water for such things as washing cars, watering lawns, or taking long showers.

"Ten liters per person per day?!" exclaims the supervisor. "Don't you think that is a bit unreasonable?"

"That is the only way we are going to get ourselves out of this mess," replies Steve. "It may seem unreasonable now, but hopefully we can save enough water to tide us over until the rain returns."

A statement dictating water conservation is issued the next day; anyone not acting in accordance would be severely fined. People are asked to use the honor system and to report anyone not abiding by the ruling. The effects of this mandate can be very easily seen: The lawns lose their lush, green color due to the lack of water. While visiting his parents, Steve notices that the grass on their front lawn still looks healthy.

"Dad, you know about the ruling that is currently in effect, right? You're not watering the lawn, are you?"

"Of course not, Steve. Would I do that to my own son? I can't explain why the grass looks so good, but who's complaining?"

Steve finds that tough to swallow, but believes his father—that is, until the following week. Still somewhat suspicious, he decides to make an unannounced visit to his parents' house, and what he finds astonishes him. When he pulls up to the house, he sees his father watering the lawn.

"Dad, what are you doing?"

"Steve, what are you doing here?"

"Dad, I can't believe you lied to me. I'm going to have to report you."

"Son, you can't do that, I'm your father. Besides that's a stupid rule. Also, do you really feel that my watering the lawn every once in a while is going to do any harm anyway?"

Questions for Discussion:

1. What are the facts in this case?

2. What would you do if you were Steve?

3. What do you think Steve will do?

4. What do you think Steve should do?

5. Is Steve's father right in what he is saying? Would Steve's choice to cover up his father's noncompliance be ethical?

6. Would his choice to report his father be *unethical*?

2.6   Case Title: Soon-to-Be-Not-So-Pleasantville (sc)
Case Type: Domestic Applications

Fact Pattern

The town of Pleasantville has had a household hazardous waste recycling program for the past few years; participation is mandatory for all of Pleasantville's
residents. This program sets requirements for the types
of waste that *must be* recycled, e.g., used oils, paints, et
cetera, and the types of waste that voluntarily *can be*
recycled. One of the wastes in the voluntary category is
used household batteries.

John, an environmental engineer for a prominent
environmental consulting company, is a resident of
Pleasantville. He follows the mandatory recycling
requirements but does not follow all the voluntary
ones. He does not recycle his used batteries because he
feels that it is too much of an inconvenience to separate
them; he already has too many garbage cans for the
various recycling programs.

John knows that Pleasantville landfills whatever
waste is not recycled. He also knows that there is a high
probability that once these batteries are landfilled, they
will leach out of the landfill and into the soil and possibly the groundwater. The problem is that these batteries
contain heavy metals such as mercury, cadmium, and
nickel, which can cause severe environmental problems.

Questions for Discussion:

1.   What are the facts in this case?

2.   Does John, an environmental engineer, have a responsibility to go above and beyond the required procedures?

2.7   Case Title: Buyer, Beware (bh)
Case Type: Domestic Applications

Fact Pattern

Erik is a former engineer who now works as a salesman for a large chemical company. The company mainly focuses on the production of chemicals for domestic applications such as household cleaners. Their newest product, Super Shine, a kitchen floor cleaner, promises to be the company's best-seller yet. Sales are through the roof, and John's hard work at promoting the product is a big reason for that success. There is scarcely a single home where Super Shine is not preferred for getting floors cleaner.

Erik likes to keep up to date on all the information available on the items he promotes. He is constantly trying to find a new and better way at making what he sells more desirable to the customer. One day while trying to find out any new pertinent information about Super Shine, he comes across a lab report marked CONFIDENTIAL. Upon reading the report, Erik discovers that Super Shine contains ingredients that are known to have adverse effects on people's health as well as cause cancer in laboratory animals. Erik, almost instantly, runs to inform his boss, Bill, of what he has found in hopes of temporarily pulling the product from the market.

"Bill, I found this lab report concerning Super Shine. We may have to pull Super Shine off the market until further testing about its safety can be completed."

"Whoa! Hold on a minute!" screams Bill. "Number one, Super Shine is our biggest seller. If we were to pull it off the market, the company would go Chapter 11, not to mention the possible lawsuits that would arise. We're not pulling anything off the market."

"I understand your concerns, Bill, but we should pursue this issue further. We have a responsibility to our customers."

"There haven't been any complaints, so I say we let the issue go, Erik. Do you want us all to be out of jobs? Trust me, Erik, everything will be fine. No one has to know."

Questions for Discussion:

1.   What are the facts in this case?

2.   What kind of problem is Erik facing?

3.   What consequences could Erik and his company face if he doesn't report what he has found?

4.   What consequences could Erik and his company face if he *does* report what he has found?

5.   Is Bill really looking out for Erik's best interests?

6.   What do you think Erik's final action will be?

> **2.8   Case Title:** Whose Car Is It, Anyway? (cp)
> **Case Type:** Domestic Applications

Fact Patterns

Randy is a junior environmental engineer at Bogart University, which is approximately a two-hour drive from his house. Randy decides he wants to bring his car back to school for the fall semester so he wouldn't be limited to the campus bus service runs and hours. He approaches his parents about his plan at dinner one night to get their approval. At first, they seem against the whole idea, but Randy manages to convince them how much easier it would be on them if they didn't have to shuttle him back and forth, a routine that forces them to make a long trip for vacations and holidays. And he promises he will drive to see his great-aunt Margaret, who lives an hour beyond his school; however, he really has no intention of visiting.

"If you're gonna take the car," his father states quite sternly, "you're gonna assume the responsibilities of having the car." Randy agrees and then goes about his activities for the rest of the evening.

In the morning, Randy's father brings him out to the driveway. Randy notices some automotive-looking parts lying on the blacktop next to a small tool bag.

"Welcome to Car Maintenance 101," his father says. "I'm gonna show you this car inside and out, just in case you ever have a problem."

Randy understands that his father's intent is to show him how to take care of the car beyond washing and waxing it and pumping the gas.

"You gotta know the mechanics of the car. Women love that. If you were a mechanical engineer instead of a tree-hugging engineer, you would know that by now," his father says in a joking grunt.

One of the drills Randy's father runs through with him is how to change the oil in the car. His father guides him through the whole process of draining the oil, changing the filter, and then finally putting in the new oil.

"That wasn't as complicated as I believed it was," Randy thinks.

As Randy wipes his hands clean on a rag, he watches with curiosity as his father walks to the backyard with the container of old oil. Randy always has wondered what his father did with old oil; as long as Randy can remember, his father has changed the oil on all five of their cars, as well as on his stepbrother's tractor trailer. To Randy's total surprise, his father begins to dump the old oil at the edge of their property. Randy knows from his various engineering courses the negative effects of backyard oil dumping.

"Dad, don't you know that by introducing a petroleum product into the environment, you are contaminating the soil, thus degrading its quality?! Not to mention the potential contamination of our water supply, since the aquifer our pump taps runs parallel to the house and ends where the elevation of the land increases near the tree line. That puts your dumpsite at around the middle of the aquifer!" Randy bursts out.

Randy's father looks at him in amazement. "I guess we're getting our money's worth from this college thing," he comments. "Those classes you take make you get a big head. The little bit of oil I throw out back here isn't going to do any harm to anything."

"You've been dumping oil here for years, to judge by your present behavior! It's not exactly a *little bit*, especially when you change the oil on the rig. The pond out in the front of the house is natural, and goes up and down with the rain. That shows you how close to the surface the top of the aquifer is. Any oil you put down is going right into our water supply," Randy argues.

"Ahhh, kid," Randy's father says with a disbelieving swat, hoping to end the conversation.

Questions for Discussion:

1.   What are the facts in this case?

2.   How would you feel if you were Randy?

3.   How would you feel if you were Randy's father after hearing what Randy has to say?

4.   What should Randy do about his father?

5.   Should Randy do anything about the potential contamination of his water?

6.   If yes, what? If not, why?

```
┌─────────────────────────────────────────────┐
│  2.9    Case Title: Don't Flush Water         │
│                 Down the Drain (robp)         │
│         Case Type: Domestic Applications      │
└─────────────────────────────────────────────┘
```

Fact Pattern

Jack is planning to remodel his bathroom. After looking at bathroom showrooms, he finally decides on a particular layout, including fixtures. Although he isn't sure if he can afford it, he has a plumber come to his house and provide him with an estimate.

The plumber quotes a price of four thousand dollars. However, he states that the toilet bowl that Jack wants cannot be installed in his bathroom.

"I'm sorry, but because you're a resident of New York state, by law I am not allowed to install the toilet that you want," says the plumber. "I can install only water-saver toilets, and that particular model you want comes only in the standard size. They don't even sell a standard toilet in New York state anymore."

Although Jack has never heard of this law, he is familiar with water-saver toilets. The bathrooms in Jack's workplace contains these toilets, and they are constantly getting stuffed up. Jack then decides to talk with his coworker Mary about his situation.

"You don't want to get a water-saver toilet," Mary comments. "Those toilets don't have enough water to force everything down the toilet."

"But it's the law," replies Jack. "And I should do my share to try to save water."

"But in reality, Jack, you will not save water because you will have to flush twice as many times as you do an ordinary toilet, thus wasting the same amount of water.

Let me give you the name and number of a plumber who will install the standard toilet for you."

Jack doesn't feel right about calling this plumber; however, he finally decides to talk with him. This plumber states that he can install a standard-size toilet as well as remodel Jack's bathroom, all for a price of thirty-five hundred dollars.

"How would you purchase a standard toilet if they don't sell them anymore?" Jack asks.

"I would go to New Jersey and buy the toilet. They still sell standard toilets in New Jersey," replies the plumber.

"You can do that?"

"Legally, no. But I won't tell anyone if you don't," comments the plumber.

Questions for Discussion:

1.   What are the facts in this case?

2.   What is the ethical problem Jack is facing?

3.   Is Mary's advice about not saving water logical?

4.   Do you think Jack's financial problem will help sway his decision to install the standard-size toilet?

5.   What do you think Jack's final decision will be?

2.10   Case Title: Help Me, Please? (ns)
Case Type: Domestic Applications

Fact Pattern

Part One

Tracie and Travis are two field technicians working for a consulting firm hired by the state DEP. Tracie is an intern with the company, and Travis is the field supervisor. They are inspecting catch basins for a statewide survey. One day, out in the field, Tracie is approached by an elderly gentleman while she is recording information.

"Excuse me, miss, do you work for the DEP?" he asks.

Tracie turns around and automatically takes notice of the old man's feeble condition and wrinkled face.

"Excuse me, what did you say?" Tracie replies.

"Do you work for the DEP? There's something wrong with my sewer connection," he repeats with somewhat greater volume. He has a soft and shaky voice that causes Tracie to squint her eyes every time he speaks, as if this can help her hear him better. She sees that he uses a nicely carved cane to walk.

"The company I work for is a subcontractor to the DEP, but I could try to help you. What's the problem?"

"Well, every time it rains, my basement overflows with sewage water. I don't understand how this happens. I have lined and relined my basement. I've hired people to come look at my connections, and they tell me nothing's wrong, yet it still continues to flood. What can you do?" he asks desperately.

"Well sir, we're just inspecting the catch basins, we're not really doing much work concerning sewer connections to houses."

"I'm eighty years old and I live alone. I am in fair health now, but I do think my basement will be the death of me," he comments as he laughs. "You know, it's spring and it rains constantly. In the past month, I have spent close to five thousand dollars just to have my basement cleaned and renovated. I have called the DEP, but they just take my complaint and tell me they'll send someone to look at it. I began calling about four months ago. Could you please look at my connections and tell me what the problem is?"

Tracie can't help but feel sympathetic toward the old man. She discusses it with Travis, who happens to be very knowledgeable about sewer connections. Luckily they have tools in the van to lift the manhole and analyze the sewer connection from the man's house. Travis tells the elderly gentleman it is possible that the sewer pipe is too small to handle the amount of waste traveling through it; thus, when there's a rainstorm, the water in the pipeline gets backed up and switches directions. Many basement lines get such backwash.

The old man is grateful that someone has given him an answer. "Well, what do I do?" he asks.

"This is not your problem. This is the state's responsibility. What I would suggest is to continue calling the DEP with your complaint," Travis answers.

"I've done that already. They don't care . . . ."

Tracie is disturbed by the old man's problem. How could the DEP ignore someone in his situation? She takes his name and number and says she'll see what she can do.

Questions for Discussion:

1. What are the facts in this case?

2. What actions can Tracie take?

3. What are the risks involved in question 2?

Part Two

The next day, Tracie calls the DEP and talks to a woman in the department where complaints are taken. She mentions the old man's name and number and asks that his file be checked. The DEP has him listed as calling eleven times.

"Why hasn't anyone helped this man?" Tracie asks angrily.

"Well, the guys who handle that problem are very busy. Things here are very backed up. There's so much that needs to be done. We'll get to it when we have time."

"But his first call was made months ago. How do you explain this?"

"Well, we're working as fast as we can. We handle the most important problems first, then we go from there."

Tracie, disappointed in her answer, abruptly says, "Thank you," and hangs up the phone. She is outraged and perplexed. She wants to help the old man and report the DEP for negligence, but she is just an intern. What good is *her* word? She also does not want to spark controversy with the DEP and the state government, or risk losing her job or jeopardize the jobs of others. However, she decides to act.

Questions for Discussion:

4. Why is Tracie in such a bad position?

5. Is taking responsibility for this man's problem a good decision for Tracie?

6. What are the possible courses of action Tracie could take?

> **2.11   Case Title:** Musical Hysteria (jns)
> **Case Type:** Domestic Applications

Fact Pattern

Part One

Every day when Eric comes home from work, he listens to his brand-new six-thousand-dollar stereo system in order to relax. It is all "hooked up," with a five-disc CD player, a 3-tape deck, a tremendous mixing board, a fifty-pound amplifier, a thirty-two-button equalizer, a microphone, six 19-inch speakers, and a subwoofer. He takes a great pride in his stereo and likes to show it off to anybody who will listen.

Eric feels that in order to get his money's worth out of the system and show it off effectively, he should use all of its magnificent qualities, including the extremely high range of volume built into the mixing board, amplifier, and subwoofer. When Eric turns the volume up, the entire house shakes and vibrates. The sound can usually be heard within a few blocks' radius from the house. His wife, Dorothy, continually tells Eric that he plays his music too loud and that she can't get any work done around the house because of it.

Eric responds, "Well, I live here, too, and I should be able to do what I want with my stuff."

Dorothy tells Eric, "I am not saying you can't listen to your stereo, just not so loud."

Since Eric just shrugs at Dorothy, she decides to leave her own house for a few hours until he is done listening, which is what she usually does when he plays his stereo.

While Dorothy is out taking a walk, her neighbor Lisa stops her to talk.

"How can you put up with all of that noise your husband makes with that stereo?" asks Lisa.

Dorothy says, "Well, it relieves his stress, so I usually just go out for a while and leave him alone."

Lisa replies, "Well, I didn't want to have to tell *you* this, but when Eric plays that stereo, my dog won't stop barking and my baby won't stop crying. It drives me crazy. I can't get anything done and I usually have a headache by the time it's all over. It is really like a nightmare when he plays it that loud."

Questions for Discussion:

1.   What are the facts in this case?

2.   Is it Dorothy's problem or Lisa's?

3.   What should Dorothy do?

4.   Who has the responsibility, Eric or Dorothy?

Part Two

Dorothy thanks Lisa for letting her know about the problem and goes home to confront Eric. She tells him all about the problem that it is causing Lisa.

Eric replies, "Well, it is not my fault Lisa can't control her dog or her baby. I can do whatever I want in my house. If I don't relax after work, I might go crazy."

Questions for Discussion:

5.   What, if anything, can Dorothy do now?

6.   What do you think her motives are for acting the way she does?

7.   Is Eric's behavior Dorothy's problem?

2.12   Case Title: Heated Arguments (rs)
Case Type: Domestic Applications

Fact Pattern

Recently, Tommy a mechanical engineer, has been looking for a new hot water heater. His cousin Frank tells
him, "My best friend owns a heating maintenance company. He could probably get you one pretty cheap."
Within the next week Frank's friend comes by to install
the hot water heater while Tommy is at work.

On his return from work, Tommy goes downstairs to
see the new heater. While looking it over, Tommy notices
that the heater does not have a safety relief valve, and
the fitting where it was to be installed is simply capped.

Concerned, Tommy calls Frank's friend, who says,
"Don't worry about it. I've never had a problem with a
heater blowing a safety valve; plus it's cheaper to get
them without the valves. All I do is tell the manufacturer
I install them myself."

Tommy knows that even a small water heater over-
pressurizing could do a lot of damage not only to a
house but also to a person. He tells Frank's friend that he
should not be installing the heaters this way, that it is
very dangerous. "Mind your own business and be grateful," Frank's friend tells him.

Not getting anywhere with this "friend," Tommy
speaks with Frank. "What your friend is doing is wrong
and extremely dangerous, Frank. You should talk to him.
If someone finds out, he could get in a heap of trouble."

"My friend knows what he's doing, Tommy. If you feel
that uncomfortable, I'll get him to get you a valve, and
then it's over. Just don't say anything else about it; do me
and him a favor."

Tommy stresses the points that someone could get killed or someone's eventually going to find out.

"Why, Tommy? Are *you* going to tell someone?" Frank asks.

Questions for Discussion:

1.   What are the facts in this case?

2.   What is Tommy's dilemma?

3.   What would you suggest Tommy do?

4.   Put yourself in Tommy's position. What do *you* do?

5.   How would you react if you were Frank or if you were Frank's friend?

> **2.13   Case Title:** Family Values (dt)
> **Case Type:** Domestic Applications

Fact Pattern

The solid waste collection agency of Tarrytown has decided to collect the garbage separated as follows: paper, plastic containers, metal cans, and garbage. This policy will enable the collection facility to make extra money by selling recyclable paper and garbage to composting facilities. In order to do that, the facility either has to sort out the mixed waste at the facility or ask the citizens to sort it at the production sites (homes). This program can help the facility in earning money to support its expenses, which would mean the community could be charged less for collection.

After this decision is made, the citizens are notified via letters and public meetings. As the notice indicates, each separate waste will be collected twice a month. Also, the facility has determined that blue and black bags are to be used for recyclable containers and garbage, respectively.

The Blackmer family has been living in Tarrytown for seven years. They think sorting the waste at home is a waste of their time. They have to spend extra money buying different-color bags, and the waste will have to be stored in their backyard for fifteen days, taking up space and causing malodorous conditions.

They decide not to recycle for the most part, thinking one noncompliant family will not hurt the facility in any way.

Questions for Discussion:

1.   What are the facts in this case?

2.   Is the Blackmer family ethical for not recycling?

3.   If many citizens would do this, what can happen to the facility?

4.   How may recycling help the Tarrytown community?

3

Health, Safety, and Accident Prevention

3.1 Case Title: The Sweet Smell of Success (ec)
Case Type: Health, Safety, and Accident
Prevention

Fact Pattern

Julie is a second-year chemist in a relatively small but competitive chemical company. She has been the assistant to senior chemist Mark, who is up for promotion this year. The two have just discovered a new chemical product, SweetSmell, which could help to reduce odor at wastewater treatment plants and landfills. Mark is ecstatic because this is the big break he has been waiting for. His company will benefit substantially from this new product, and he undoubtedly will get the promotion he has wanted for so long.

Julie too is very happy with their discovery. However, as a young, eager, and relatively new employee, she feels obliged to investigate their new product further. To her surprise, she finds the chemical could have an adverse impact on humans if it gets into their drinking water supply.

"Mark, I have bad news. We cannot introduce Sweet-Smell at the next meeting. I have discovered it can have harmful effects on humans."

"You do not know what you are talking about, Julie. SweetSmell is not harmful to anything or anyone."

"But, Mark, I can prove it," proclaims Julie.

"Listen to me. This is my big break, and I will not have anyone ruin it. Stay quiet, or I will have you fired by tomorrow morning."

Questions for Discussion:

1. What are the facts in this case?

2. What is the situation Julie is faced with?

3. Are Mark's actions ethical?

4. What do you think Julie will do?

5. What *should* Julie do?

> ### 3.2 Case Title: Danger at Home (bd)
> ### Case Type: Health, Safety, and
> ###   Accident Prevention

Fact Pattern

Mike, an engineer, is retained to investigate the structural integrity of a fifty-year-old occupied apartment building that his client is planning to sell. The structural report written by Mike is to remain confidential under the terms of the agreement with the client. In addition, the client makes clear that the building is being sold "as is," and he is not planning to take any remedial action to repair or alter any system within the building prior to its sale.

Mike performs various structural tests on the building and reaches the determination that the building is structurally sound. During the course of performing services, however, the client confides to Mike that the building contains inadequacies in the electrical and mechanical systems. These inadequacies pose a violation to pertinent codes and standards. Even though Mike is not an electrical or mechanical engineer, he does know that those inadequacies could cause injury to the tenants of the building, and he informs the client accordingly.

In his report, Mike briefly mentions his conversation with the client concerning the building's deficiencies. Due to the terms of the agreement, however, Mike decides not to report the safety violations to any third party.

Questions for Discussion:

1. What are the facts in this case?

2. What are the possible courses of action Mike could take?

3. Do you think it is ethical for Mike not to report the safety violations to the appropriate public authorities?

4. Do you feel that Mike's duty—as a professional engineer—to hold public health, safety, and welfare paramount outweighs any other consideration?

3.3 Case Title: Medicine versus the
Environment (bd)
Case Type: Health, Safety, and Accident
Prevention

Fact Pattern

Carol is facing a medical dilemma not likely to be solved during her lifetime:

As a member of the Audubon Society and the Nature Conservancy, she has actively embraced the conservation movement and led children on nature hikes.

But Carol, now sixty-five years old, also is battling ovarian cancer and wants an experimental drug called Taxol. The drug, derived from the rare Pacific yew tree, has achieved a 30 to 40 percent response rate in advanced cases of ovarian cancer. Medical researchers say the only way to produce enough Taxol needed for treatment and research would be to chop down hundreds of thousands of yews, which are a refuge for the spotted owl.

"I want to be treated," Carol says, quite apologetically. "However, I don't want to see the forest destroyed."

The conservationists say that they are concerned about saving the owl and the yew trees as part of their overall mission to preserve the diversity and integrity of threatened forests. Conservationists, of course, deny that they are inhumane. They also say that medical researchers and a major drug company that manufactures Taxol aren't moving quickly enough in pursuing other options for making the drug. After all, the environmentalists strongly believe that the ancient forests that gave us the yew may provide us with answers to other medical problems we haven't thought about yet; are people enti-

tled to destroy all of nature for their own selfish purposes? they ask.

Clearly, this is part of the ultimate confrontation
between medicine and the environment.

Questions for Discussion:

1.   What are the facts in this case?

2.   The environmentalists insist they have patients' long-
term interests at heart. Do you agree?

3.   Balancing finite resources with finite lives is the challenge now facing medical researchers. Does one concern
outweigh the other?

4.   What are the possible courses of action the researchers could take?

> **3.4 Case Title:** An Offer I Can't Refuse (gd)
> **Case Type:** Health, Safety, and Accident
> Prevention

Fact Pattern

Joseph is a field technician for a chemistry laboratory. His job today is to go to P.S. 1000, take lead wipe samples from some of the classrooms, and test them for any lead contamination. The school needs these samples taken so that they can open for instruction the next day. Joseph gathers up his equipment and heads off to the school; when he arrives he is immediately greeted by a member of the school's staff.

"Hello there, sir. I am the head of the school administration and in charge of making sure that the school reopens tomorrow."

Startled by the directness of the administrator, Joseph cautiously replies, "Hi, how are you doing today?"

"Good, good, but I'd be a lot better if this whole process were over with and knew the school would reopen tomorrow," says the administrator, who is standing in the doorway to the school.

"Well, let's get started, then. If you will let me get to the classrooms, I just might be able to take the lead wipe samples," responds Joseph somewhat bluntly.

"Son, there is just one thing I have to say to you first. If this school is not reopened tomorrow, a lot of parents will be very angry with me. Do you want that to happen?"

"Sir, that is none of my business; I just take the samples," says Joseph, who is growing more annoyed with the administrator.

"OK, let me make myself a little clearer. The school *will* reopen tomorrow, right?" the school representative asks as he slips Joseph a fifty-dollar bill.

Startled by this, Joseph says, "I am sorry, but I don't accept bribes. Take it back!"

The even more insistent administrator answers, "Look, just take it and think about what I told you. If you don't, I don't want to imagine what will happen to you. Do you understand what I am saying?"

Joseph steps back for a second and asks the administrator, "Are you threatening me?"

The administrator turns around, starts to walk away, and says, "You know what you have to do."

Questions for Discussion:

1. What are the facts in this case?

2. Why would the administrator bribe Joseph?

3. Should Joseph accept the money and pass the samples?

4. Should Joseph tell the proper authorities about what has just happened?

5. Should he just do his job properly and not worry about the administrator's threat?

6. Do you think the administrator will harm Joseph if the samples fail?

7. What might happen to Joseph if he is caught taking a bribe?

8. Could the "administrator" be an undercover detective wanting to see what Joseph will do?

> **3.5   Case Title:** Why Wait? (pf)
> **Case Type:** Health, Safety, and
> Accident Prevention

Fact Pattern

Dr. Kevin Inten and Dr. Whitney Bugh are environmental chemistry professors at a small university; since the time they met each other in graduate school over thirty years ago, they have worked together and been good friends. Their efforts in the development of methods for the handling and cleanup of toxic substances are world renowned. Dr. Inten has been hired as a consultant for the hazardous substances division of a major chemical company.

Dr. Inten has requested that the company hire Dr. Bugh to assist him in the work he has been assigned. The company agrees, and requests that Drs. Bugh and Inten perform a series of tests on the chemicals stored in this division and determine new methods of disposing of them safely.

For quite some time Dr. Bugh has been requesting that her school purchase several chemicals needed for her graduate students to continue their research. The university has informed her that they cannot afford to purchase the chemicals anymore, since the price of the chemicals has risen due to their recent designation as "mildly toxic." Dr. Bugh has asked the company if they might be able to donate some of their supply of these chemicals to the school so that the students can finish their research and graduate on time; the company manager approves the chemical's donation. The only problem is that the processing time for the necessary paperwork is more than six months. Dr. Bugh's students are scheduled to graduate in three months, and extending their research for an additional three months will not be possible. She mulls over the situation.

While working on her assignment, she walks into a back stockroom of the hazardous substances division and discovers the chemicals her students need for their research. Part of her responsibility has been to perform, with Dr. Inten, an inventory of the types and amounts of chemicals that are stored in this division of the company. Dr. Bugh realizes that no one would know that some of the chemicals were missing if she were to take them and not report these amounts in her evaluation. As Dr. Bugh is placing some of the carefully sealed containers holding the chemicals in her bag before leaving one night, she is spotted by Dr. Inten, who proceeds to ask her what's going on.

"Whitney, please tell me you're not doing what I think you're doing," says Dr. Inten.

"Relax, Kevin, I *know* what I'm doing. My students need these chemicals in order to finish their research and graduate on time. Besides, I'm really not doing anything wrong. The company said I could have the chemicals; I just can't afford to wait that long to get them because some bureaucrats want to put their stamp on a piece of paper," replies Dr. Bugh.

"But you're stealing and could get in a lot of trouble if you're caught with those chemicals because they're toxic," Dr. Inten comments.

"It's not really stealing. They wanted us to figure out ways to get rid of these chemicals and weren't going to use them anyway. Plus, these chemicals are only mildly toxic, and you know I have been handling them safely for years. Now, can you help me carry some of them?" asks Dr. Bugh.

"Sorry, I have to go right now. I promised I would pick my wife up and I can't be late," responds Dr. Inten as he hurries out the door.

Dr. Inten is troubled by what his friend is doing and has made up that excuse to leave quickly because he doesn't know what to do. On one hand, Dr. Bugh is right: The company isn't going to use the chemicals, and they have said they would give them to her eventually. She is just trying to help her students graduate on time, and she *is* an expert on handling these types of chemicals, Dr. Inten rationalizes. Yet, he knows the company's policy on removing chemicals from the workplace. He anticipates that he will be fired if it is discovered that he knows what Dr. Bugh is doing; however, he chooses not to report her. Reporting her would mean ruining the reputation of his friend, who believes she is doing the best thing, under the circumstances. Their families are close, and Dr. Inten believes he might lose her friendship and, further, the trust of his academic colleagues for reporting a friend.

These thoughts race through Dr. Inten's head as he drives home, torn by his conflicting feelings about what is the correct thing to do.

Questions for Discussion:

1. What are the facts in this case?

2. What ethical dilemma has occurred?

3. Is Dr. Bugh correct or wrong in doing what she has chosen to do?

4. What are some of the positive or negative ramifications of Dr. Bugh's actions?

5. What are some of the ramifications of the decision Dr. Inten must make?

6. What choice do you believe Dr. Inten will make and why?

3.6   Case Title: Likely Stroke Cigarette
Company (sf)
Case Type: Health, Safety, and Accident
Prevention

Fact Pattern

Introduction
It may seem surprising to many, given what we now
know about the dangers of tobacco products, that manufacturers of tobacco products thoroughly test their products, not just for quality control, but also for all the same
reasons that independent labs conduct tobacco testing on
behalf of lawyers and in the process of routine cancer
research. Additives are usually quite benign if inhaled
separately, but in the cone of intense heat at the tip of a
cigarette being "dragged" on, reactions may take place
between additives and the ten thousand–odd natural
plant products in the tobacco matrix. Cigarette paper
may also be a source of toxins when it is combusted; for
instance, the carcinogen benzo[a]pyrene, a polycyclic
aromatic hydrocarbon, is a combustion product of
tobacco and paper. The carcinogenic N-nitrosamines are
naturally present, and some are formed during combustion from nicotine and from amino acids that are naturally present.

Part One
Product development researchers at Likely Stroke Inc.
(LS) would like to introduce to their cigarettes a new flavor additive that smells like mint; they feel it would
greatly enhance the quality of the smoke, and very little
is required for olfactory detection. Joan, who works in
research and development, is asked to do a preliminary

inhalation toxicity study on mice using varying levels of the additive in an already marketed product. Several weeks later, after the mice have been ordered, the cigarettes spiked with different amounts of the new additive and conditioned, and the apparatus set up, testing begins on four groups of fifty mice. The mice, each in an individual chamber, are given smoke at a rate commensurate with their body weight, roughly equivalent to smoking thirty cigarettes/day under standard smoking conditions (the smoke machine drags 35 microliters for 1.5 seconds/puff, one time/minute).

The study is carried out for a duration of ten weeks, during which time four of the two hundred mice in the study expire, and six develop precancerous lesions in the mouth and throat. No trend exists in the mortality rate, which is in fact typical among mice. The cancerous lesions occur in all groups, but the two groups with the two highest amounts of the flavor additive each generate two mice with cancer; the two groups with lower amounts of additive, each one mouse with cancer. This would seemingly point to a trend; however, even if all four groups had experienced the exact same conditions, there is a significant chance this type of distribution would have occurred anyway.

So there is little indication that the additive makes the product more carcinogenic. If there were several more cancers in proportion to the amount of additive, this would indicate a definite need to enlarge the study.

Questions for Discussion:

1. What are the facts in this case?

2. Does the additive pose an increased risk of cancer?

3. Can this study be considered conclusive?

Part Two

Meanwhile, Tom in corporate analytical has been given the routine task of analyzing the smoke from cigarettes of the same batch used in the animal testing (i.e., random samples are chosen from each group prepared for the mice). The amounts of tar and nicotine are all well within the normal ranges, as are the usual nitrosamines, but there is a change in the relative amounts of some nitrosamines, which correspond to the amount of additive.

It is necessary to give some background into how this sort of testing is done. Cigarettes are smoked on a machine under the industry standard conditions mentioned in part one of this case study, and the smoke condensate is collected on a filter in the smoke machine. The filter is then placed in solvent, and the chemicals to be tested for "extract" into the solvent. Most of the solvent is then removed, and a chemical called an internal standard is added. The sample is then diluted back to a specified volume and is ready for analysis.

Looking for specific chemicals in cigarette smoke condensate is tantamount to looking for a "needle in a haystack" because there are so many similar compounds that they are extremely difficult to separate cleanly. Looking for one kind in particular requires special instrumentation. To test for nitrosamines, a few microliters of the sample is injected into an instrument called a gas chromatograph. This can be a simple, long thin tube called a column packed with fine-grained silica powder (easy to make and cheap). Or it can be an open capillary column, a tube whose inside walls are coated with silica in what is called a "bonded phase" (hard to make and expensive). The tube has a heated injector (to vaporize the sample in the gas phase) at one end, where the sample is injected, and a detector at the other end, which gives a response linearly proportional to the amount of that type of chemical present. The column is run through an oven and is kept hot so that all of the compounds in the sample will remain in the vapor phase and all—it is hoped—will make it out the other end eventually. Chemicals are separated in the column (mostly in order of increasing boiling point) so that they tend to emerge from the column only one or a few at a time, and conditions are adjusted so that the best or cleanest separation takes place between the chemicals for which you are analyzing. The result is a chromatogram, which is simply a chart of detector response versus time, with chemical detection in the form of "peaks," and with certain chemicals emerging at specific times relative to when the internal standard emerges.

The choice of detector is critical to the specific types of chemicals for which you wish to analyze: Some detectors will pick up everything and the output will look like noise, and some detectors will pick up only certain types of chemicals. The latter type is preferred, and in the case of nitrosamines, the thermal energy analyzer/detector is the only way to pick them out of the soup of ten thousand–odd ingredients (imagine looking for five to seven chemicals out of this many). Because smoke condensate is filthy, the cheaper, packed column is invariably used to analyze it. Generally it gives adequate separation of the nitrosamines.

Now, back to the story. As already mentioned, there are five to seven kinds of nitrosamine normally present in tobacco smoke condensate. The internal standard is a nitrosamine that does not occur in tobacco or its smoke, but is used to confirm the emergence time and quantify the amounts of the nitrosamines in the smoke condensate samples. The curious effect of the additive is that one of these peaks, corresponding to a particularly toxic cancer-causing nitrosamine called NNK, has decreased in size, roughly in proportion to the amount of additive in the cigarette. An increase in the size of a peak corresponding to a less cancerous nitrosamine called NAT is also observed, also in roughly the same amounts. This is great! Could it be an additive that improves the flavor of the product *and* makes it less harmful?

Questions for Discussion:

4. What are the facts in this case?

5. Because tobacco is a natural product, the content of each type of nitrosamine varies with each crop or even each plant in a crop. Is such variation relevant to this case?

Part Three
A meeting has been scheduled between the head of product development and the research staff to discuss the results of the various studies under way on this additive. Tom and Joan present their findings, which basically show that there is little conclusive evidence of increased cancer or any other kind of risk resulting directly from the new additive, pending further study.

Tom is truly an expert at chromatography, and in his presentation he discusses the fact that along with the increase in the size of the NAT peak, another effect is noticed: The height/width of the peak is somewhat lower than it should be for a compound emerging from the column in only fifteen minutes. (Peak broadening, as it is called, increases proportionally with the time it takes for a chemical to emerge from a column). This is slightly disturbing because it means that there may in fact be two compounds emerging from the column at very nearly the same time. Also, it occurs only in the samples containing the additive. If this is the case, then a new type of nitrosamine may be present, and it should be investigated. Unfortunately, separating NAT from another possible compound with a very similar emergence time on a packed column is impossible, and it may not be worth ruining an expensive capillary column in very short order to investigate another compound that may not even be present. Joe, the head of the division, has concluded that further study is not necessary, and LS will go ahead and use the additive.

Questions for Discussion:

6.   What are the facts at this point?

7.   Since very little evidence exists at this point of significant risk, is it worth investigating this new compound, from a management standpoint?

Part Four
Although Tom is very busy, he decides to run samples from all the smoke tests, starting with a nonadditive sample followed by one taken from the test containing the most additive, through a capillary column that is dying anyway. The result is that the samples from the additive cigarettes indeed contain two peaks, (although they still overlap somewhat), whereas there had been only the one for NAT. This means that there is indeed a new compound formed. The column held up only long enough to inject two more samples—the two intermediate ones—thereby roughly establishing a trend.

Questions for Discussion:

8.   What are the facts at this point?

9.   Should Tom inform Joe about the confirmation of the presence of a new nitrosamine?

10.   Because capillary columns are so expensive, most independent labs are not likely to repeat this experiment; this new chemical would probably not be detected by them. From a legal standpoint, is it worth determining what this new chemical may be and investigating its properties?

Part Five

Tom has logged his finding in his lab notebook and has decided to tell Joe about it when Joe returns from vacation in Vegas. Meanwhile, he is very excited about his find, so he chooses to tell one other person: his friend Sue, a synthetic organic chemist from another department with whom he frequently has lunch. (It should be noted that in most corporations, particularly those for whom public relations may involve sensitive issues, there are strict rules about discussion of work matters outside the supervisor/subordinate/group relationship.) Tom asks Sue to have lunch with him at a restaurant not usually frequented by employees of LS.

She concludes that there may indeed be a reaction occurring between the additive and one or more components within the tobacco matrix, and that the new compound must have a very similar structure to NAT. Because she has synthesized NAT in the past, she decides to check her notes to see what side products may have been formed when she last made it, what similar N-nitrosamine chemicals may exist, and how they could be made if not already available.

A week later, a very depressed Joe is back from his vacation ("Viva Lost Wages," as the saying goes). Tom decides to give him some space for a while. Meanwhile, Sue has surmised three or four possible identities for the new chemical, and two can be easily made. A third possibility may be the same chemical as an impurity or side product that occurs in small amounts in the synthesis of

NNK but gets removed in purification. Tom injects a solution containing only purified NAT and the unpurified NNK in a capillary column and finds that one impurity in the NAT solution has the same emergence time as the mystery chemical.

Questions for Discussion:

11.   What are the new facts at this point?

12.   Exact emergence times in gas chromatography are strong but not conclusive confirmation of the identity of a compound. Should Tom wait until he has tested the other possibilities before informing Joe about his findings?

Part Six

Tom has decided to wait before telling Joe; so he tests the other possible chemicals. None of them have the same emergence time; he concludes that the mystery chemical is most likely to be the same as the impurity from the laboratory synthesis of NNK. Sue has recently characterized this to be the nitrosamine NNAL. It does not normally occur in tobacco or tobacco smoke, and its properties have not been studied.

Tom discloses his findings to Joe and gets a lecture about responsibility and proper conduct in the workplace. Joe is pretending to be angry that any time was spent on this study after he had given the final order on this additive, which has been very successful in capturing more market share for LS. But, Joe's real concerns are that Tom has unwittingly created more work for him and that this compound's possible adverse effects could cause LS to yank the product, which has put Joe in line for a promotion. Now the identity of this chemical must be confirmed, and toxicity and carcinogenicity data collected.

Joe realizes that he is being unfair to Tom and that he should really be glad to have such a diligent employee, so after apologizing to Tom for the unnecessary lecture on corporate conduct, he explains to Tom that this study will be continued by the department of toxicology at some later date and that Tom should now forget about the matter.

Questions for Discussion:

13.   What are the facts at this point?

14.   Should Tom now let the matter drop?

Part Seven
Tom chalks his discovery up to a nice bit of chemical investigation and turns his attention to other analyses. Meanwhile, the toxicology department performs the usual battery of tests and confirms the identity of the chemical. NNAL has also been determined to be nearly fifty times as carcinogenic as NNK. This data is kept secret. Ten years later an independent research foundation reaches similar conclusions and publishes the findings in the *Journal of Carcinogenesis*.

Questions for Discussion:

15.   What are the facts now?

16.   Should LS voluntarily yank the product because of the presence of NNAL in the smoke?

17.   There is little empirical data that this flavored cigarette is responsible for more deaths and illnesses than any other similar product now on the market. What would you do if you were the decision maker at LS?

Part Eight

A consumer watchdog group and a state government have filed a lawsuit to recover medical costs from LS on the basis of the finding that NNAL is a cancerous component unique to this product. LS at first denies having any knowledge the NNAL exists in their product, buying some time to craft a defense. The CEO of LS has decided that it can tie up the courts for a few more years by claiming that LS itself will carry out the same study, and then it will simply dispute the findings of the research foundation.

Questions for Discussion:

18.   Is this ethical behavior on the part of LS?

19.   This may not be unusual behavior for an industry that has such an enormous profit margin. Besides, tobacco products contain numerous carcinogens, many equally as potent as NNAL, so why should it be unique in determining the fate of the product, which has been a huge success worldwide for LS?

3.7   Case Title: What Would Jack
Nicklaus Think? (bh)
Case Type: Health, Safety, and
Accident Prevention

Fact Pattern

Tom and the consulting firm he works for are in the very profitable business of converting old landfills into championship golf courses. They accomplish this by building a leachate collection system below the landfill in order to collect and treat waste before it enters the water table. Then, when the landfill is at maximum capacity, they seal it with a clay cap and top it with grassy vegetation. It is Tom's job to design the leachate collection systems as well as to monitor the progress of any of the firm's preexisting projects. Tom and the firm pride themselves on the fact that they provide an efficient and safe method of solid waste management in addition to an aesthetic final product.

One of the firm's most profitable preexisting projects is the beautiful Theodore Country Club. This once disgusting and toxic landfill is now the future home of the U.S. Open golf championship. Tom routinely checks the former landfill's leachate system for the "unlikely" possibility of leaks and also manages simultaneously to get in a couple of holes of golf. The aquifer below the landfill provides the community of about two thousand persons who live near the country club with drinking water. Any contamination would have catastrophic effects on the community and on the profits of Tom's consulting firm. The leachate collection system under the club is not new and has begun to show signs that it might be deteriorating.

Tom, a typical engineer, has begun to doubt his initial design for the leachate system, feeling that he can improve on its safety. If Tom attempts a new method of inserting a compacted clay liner below the collection layer, the possibility of groundwater contamination could be almost eliminated. There is only one catch to this idea: The beautiful Theodore Country Club would have to be torn up and redone, a process that could take years to complete. Regardless of this fact, Tom feels that this is a necessary precaution and decides to present his proposal to his associates and the board members of the country club.

"This is an impossibility," barks the president of the club. "We plan on hosting the U.S. Open here in three years. What would Jack Nicklaus think if his drive wound up in a thirty-yard ditch in the middle of the fairway? The club and the course would be ruined."

"Not to mention, Tom, *we* would be ruined," charges one of the firm's associates. "We could never handle the financial burden such a project would create. Our whole reputation would be compromised if word got out about this. Anyway, the leachate systems are built to last."

"Right, but for how long?" Tom replies. "There is a very real possibility that the collection system will fail within the next two years. Then what are we going to do?"

"We'll worry about that when and if it ever happens. For now, scrap the idea and focus on your current projects."

"Yes, Thomas, well done, but the club just can't afford to lose the Open."

"Thank you for your time anyway, gentlemen," Tom says bitterly. "I'd still like you to consider it a little further when you can, please."

"Right, Tom, when we get a chance. Thank you."

Questions for Discussion:

1. What are the facts in this case?

2. What type of problem is Tom facing?

3. What could be the consequences if Tom, the consulting firm, and the country club board of trustees don't correct this problem?

4. What could be the consequences if Tom, the consulting firm, and the country club board of trustees *do* correct this problem?

5. Whose best interests are the firm and the board looking out for, the community's or their own?

6. Say that the leachate system *does* last long enough to make it to the U.S. Open, do you think the firm and the country club will reconsider Tom's proposal afterward?

7. What do you think Tom's final decision might be?

3.8   Case Title: The Spill (sm)
Case Type: Health, Safety, and
Accident Prevention

Fact Pattern

Kate has just been hired by the Dewey Howe & Cheatem Chemical Company (DHC) as an apprentice at one of their manufacturing facilities. For her first assignment, she has been asked to move a container that is out on the production floor into the store room. The drum is labeled with a chemical Chemical Abstract Service (CAS) number and a National Fire Protection Association (NFPA) hazard diamond per state and federal regulations.

As she is taking the drum over to the store room, the hand truck she is using tips over, and the drum falls off. The chemical spills onto the floor and her legs, and Kate gets severely burned.

DHC says that Kate should have known what was in the drum and what precautions to take while transporting the drum. Kate replies that the drum should have been labeled more clearly, with a clear statement about its contents. DHC indicates that they are not going to cover any expenses accrued by Kate, since she was negligent in her handling of the drum.

Questions for Discussion:

1.   What are the facts in this case?

2.   Is Kate correct in handling a drum with contents unknown to her?

3. Is Dewey Howe & Cheatem correct in stating that Kate was negligent?

4. Should Dewey Howe & Cheatem have provided Kate with some training regarding the handling of these drums?

> **3.9   Case Title:** Make Way for Progress (cp)
> **Case Type:** Health, Safety, and Accident
> Prevention

Fact Pattern

Eddie has been employed in an aluminum can producing plant for about one year. The longer an employee works there, the more privileges and overtime he receives.

Eddie has made friends with Larry, who works in the front end of the plant running the "punching press," an enormous machine that cuts a circular piece of aluminum from a very large, very heavy roll of aluminum.

This plant and others like it run on very tight schedules. The workers must produce a certain number of cans by the end of the shift; if they don't, the following shifts must try to compensate, within reason. The plant managers strive not only for efficiency but for safety. Eddie, Larry, and all other employees take a safety training course every year that reviews the Occupational Safety and Health Administration (OSHA) safety standards and regulations that apply to the plant. It can be said the course is quite effective: Despite the many possible injuries that can result from working with heavy machinery, only a few cuts and bruises have occurred.

One day, members of the crew that Eddie and Larry are on find that the previous crew had some major mechanical problems with a vital machine. Because of this, the crew fell behind in production. Eddie's crew has a record of surpassing quotas or at least meeting them. Since the previous crew fell so far behind, Eddie's crew is offered a bonus if they can get production back on balance. Eddie and the rest of the crew are quite excited and go to work with the common goal of getting the bonus.

At some point during the shift, Larry's machine begins to get close to the end of a spool of aluminum. Larry begins to look around for the forklift driver but sees none. Larry knows that if he leaves his area, he will have to shut down the machine, which in turn will slow down the entire line. Just as Larry is about to use up his last hundred feet of aluminum, Eddie comes walking by.

"Eddie, where is the forklift driver?" Larry yells over the factory's noise.

"I think he is at lunch already," Eddie answers, just noticing Larry's problem. "Gimme the keys to the forklift, I'll load you up."

"No way, I know you don't have a license for the forklift yet; if you get caught driving it, we're both in a lot of trouble for breaking the rules!" Larry shouts.

"Look, if you have to shut down you machine to wait for the forklift driver to come back, there's no way we're gonna get that bonus. Now gimme the key!" Eddie yells.

Larry knows Eddie is right, but he also knows what he was taught during safety training. But Larry really could use the money for a down payment on the new truck he wants to buy. And Eddie *has* driven the forklifts illegally before, but never with Larry's key.

Questions for Discussion:

1. What are the facts in this case?

2. What should Larry do?

3. Is the risk Eddie's going to take worth the possible bonus?

4. Would you drive the forklift if you were Eddie?

5. What else could Larry do in this situation?

3.10   Case Title: Safety Comes First (robp)
Case Type: Health, Safety, and
Accident Prevention

Fact Pattern

After several men have been injured in a cogeneration facility, the plant manager decides to establish a safety committee, which consists of the operations manager and maintenance manager as well as three technicians. Their ultimate goal is to find and eliminate all the safety hazards that exist in this facility. As an incentive, the plant manager decides to reward this committee monetarily whenever a given year passes with the number of injuries being lower than the previous year.

In the course of several years, the committee has eliminated many of the safety hazards in the facility; concomitantly, the cost of workers' compensation has decreased considerably, and the number of injured people per year has reached an all-time low. Both the plant manager and coworkers are very happy with the results generated by this committee.

Although the committee has been solely responsible for the increase in safety awareness at the facility, the plant manager receives all the recognition and praise without acknowledging them; he also has not provided any of the promised monetary incentive to them.

The safety committee has become very angry and resentful. "Let's forget about using the allotted safety budget for safety; let's take some of the money for ourselves. We deserve it. And anyway, we were promised some form of monetary incentive!" shouts one of the members.

"Wait! We can't do that. Our coworkers depend on us to use those funds to help eliminate the safety hazards that still exist and to provide safety awareness courses," explains the operations manager.

Questions for Discussion:

1.   What are the facts in this case?

2.   Do you think that the operations manager's comment will affect anyone in the committee?

3.   What dilemma is the committee facing?

4.   What are the two things that the plant manager failed to give the safety committee?

5.   How do you think most of the members in the committee feel about what is happening?

6.   What do you think this committee should do?

> ### 3.11 Case Title: The Weak Link (jns)
> Case Type: Health, Safety, and
> Accident Prevention

Fact Pattern

Dave just started his job as an oiler in the engine room on
a steam-powered freighter that exports raw materials to
Europe. One of Dave's job responsibilities is to make
rounds and see that everything in the engine room is
functioning properly. All reports are written in a log
book.

Late in the day, just before Dave is ready to report that
everything went OK, he goes to his last stop to check the
water filtration system. The water from the ocean and
water from the ship that has already been used gets fil-
tered and recycled through the water filtration unit so it
can be used again as clean water. On most ships, the efficiency of this unit is very high, and the recycled water is
usually better than water in most people's homes.

As Dave is checking the filter system, he notices that
the recycled water looks a little thick and has a slight
color to it. He immediately sees this as a big problem,
and is confused about why it hasn't already been taken
care of and why he hasn't been informed by the oilers in
charge of running and regulating the unit.

"They couldn't have overlooked something like this,"
Dave thinks to himself.

With a head full of questions demanding explanations,
he asks the other oilers previously on watch if they
noticed that the recycled water looked a little brown;
each one denies any knowledge of such a problem. Since
he can't prove that any of the other oilers knew of the
problem, Dave decides that it is his responsibility to do

something about it. His next task is to go to his watch supervisor, the First Assistant Engineer, who tells Dave that it isn't his problem to deal with, but that he will tell his superior, the Chief Engineer, about the problem.

When the First Assistant Engineer goes to see the Chief Engineer, he is told to make an appointment. Two days pass, and Dave is getting agitated. He returns to the First Assistant Engineer because he doesn't want to break the chain of command, and asks him what happened.

"The Chief Engineer told me he'd get back to me, so it's out of my hands," says the First Assistant Engineer.

"Sir, what do you mean, it's out of your hands, sir? Sir, how are you going to wash those hands if the water remains dirty or gets worse, sir?" retorts Dave.

"I won't have you talk to me like that, and I don't want you telling anyone else about the filtration system. It's in the hands of the higher officers now, so just forget it," concludes the First Assistant Engineer, who then walks away.

Dave will *not* just forget this, though. He thinks it a little strange that no one else really cares that their drinking water might be a health risk. Therefore, Dave decides to put his job on the line and go straight to the Chief Engineer himself, and if that doesn't work, he will go to the Captain.

The Chief Engineer sits Dave down and tells him, "I'm going to overlook the fact that you are totally out of line and tell you that the water you think is dirty is not really that bad. The Captain and I know that there is a problem with the filtration unit, but we don't want to get everyone all hysterical about it. We are on a schedule, and things of greater importance have to get done first."

"Sir, but what about everyone's health, sir? Sir, you can't just write it off to the lack of ability of officers to do their jobs correctly and let them cover up for faulty equipment, sir," blurts Dave.

"We're working on the problem. Let us handle it. Remember what you are and where you are and who your superiors are. It would be a shame if your insubordination was logged in the log book and you were fired at the next port. There are plenty of other people willing to do your job."

With that, the oiler is dismissed for the day.

Questions for Discussion:

1. What are the facts in this case?

2. Who is responsible for the faulty equipment: the oilers, the First Assistant Engineer, the Chief Engineer, or the Captain?

3. Is this Dave's problem?

4. Is Dave acting out of line?

5. What are Dave's options?

6. What are the risks if Dave acts further?

7. What are the risks if Dave *doesn't* act further but instead keeps quiet?

3.12   Case Title: Space Exploration (rs)
Case Type: Health, Safety, and
Accident Prevention

Fact Pattern

As a new safety engineer for a utility, Michela is given responsibility for developing a corporate confined-space procedure as mandated by OSHA regulations. During her research Michela reads that the definition of a confined space is one that has limited means of ingress and egress, and is not designed for normal human occupation.

Taking time to tour the individual plants owned and operated by her company, Michela notices that most of the plants have a pump pit, which is located below the deaerator (equipment used to scavenge oxygen from feedwater prior to its introduction to a boiler). These pumps pits are generally tight areas, serviced by one or two ladders, which house a number of feedpumps.

In speaking with the plant chief operators, Michela finds that most operators have to make frequent trips into the pump pit to initiate or stop pumps. Additionally, members of the maintenance crew perform routine maintenance on the pumps in the pit. None of the chief operators considers the pit a confined space. Questioning *why* the pit is not considered one, Michela gets the following responses:

1. There are two ladders into the pit; there's no limited access.

2. Since the operators make frequent trips into the pit, it would be a hassle to go through the process for confined-space entry each time someone went in.

3. We discussed it at the safety meeting. There are no real hazards in the pit, and we feel it does not meet the definition of a confined space.

4. The pump pit is a bottom floor. It is another room in the plant, so it is designed for normal occupation.

Concerned about the possibility of the pits being a confined space, Michela requests an official interpretation from OSHA. The reply is, "OSHA cannot define specific confined spaces; it is up to the judgment of the employer to do so. If hazards, potential or actual, are present in the area, then it is a confined space."

Michela ponders, "The high-pressure water in the pit is a hazard if something goes wrong, but we have high-pressure steam pipes in the plant. It's possible that if we have a CFC [chlorofluorocarbon, a refrigerant] leak from the chiller, it will find its way into the pit and displace the oxygen. There might even be other potential hazards."

Taking a tour around the plants again, Michela notices that three of the four pits have chillers and CFC storage nearby. All four plants also have chemical storage nearby. "I have to make a decision," Michela says to herself.

Questions for Discussion:

1. What are the facts in this case?

2. What is the dilemma that Michela is facing?

3. What are some of the effects of declaring the pits a confined space?

4. What are some of the potential effects of *not* declaring the pits a confined space? In your opinion is the convenience of the status quo worth the risks?

5. What decision would you make? Why?

3.13   Case Title: OSHA Amateurs (dt)
Case Type: Health, Safety, and
Accident Prevention

Fact Pattern

Bill recently has accepted a job at an environmental chemistry lab where various kinds of toxic and nontoxic chemicals, gases, and solutions are used. The lab is located in a closed space with limited ventilation.

After Bill starts his new job, one of his coworkers, John, has to deal with an acid spillage. His colleagues are unable to locate specific neutralizers and ventilation equipment. This incident does not cause any physical harm but creates much commotion and leads to debates and conflicts among the lab crew. Many of the department workers then organize an unofficial OSHA training session without the permission of the lab authority, but not all workers are involved in this training.

Bill, who had little health, safety, and accident prevention instruction in his primary training, takes matters into his own hands, sorting out the accident details and planning a meeting with the lab manager. He wants to propose mandatory OSHA training and the procurement of equipment for safety.

The next day Bill talks to his friend Herry, who says, "Bill, you are taking things way too seriously. I don't think you should worry about this. Also, as a new employee, you don't even know the managers. I suggest you should close the case right here and not utter another word about it if you want to keep your job." Bill tries to present reasons for his intervention in the name of safety, but Herry denies it's a good idea.

Bill finds himself in a predicament. He has two options: One, go to the meeting and discuss the proposal; or, two, keep quiet. Both alternatives could lead to various parties being hurt physically or financially. Any lack of action could mean his coworkers and he himself will remain in jeopardy. But acting on the safety issues may put his job at risk. It can also bring a lot of trouble to the company if the government authorities inspect the lab. Why is it only *his* responsibility to notify them of the problems? he wonders. Should he go with the proposal or another alternative? He certainly does not have as much support on his proposal as he would like, so he frets about his next move.

Questions for Discussion:

1. What are the facts in this case?

2. How can Herry's advice affect Bill's decision?

3. Should the lab just be equipped properly, or should OSHA training also be required?

4. Is Herry's point of view an ethical view?

5. What should Bill do? Why?

3.14   Case Title: Caught in the Act (mz)
Case Type: Health, Safety, and
Accident Prevention

Fact Pattern

ACE Enviro Services has been contracted to clean up
hazardous solvents dumped from a dry cleaning business; the solvents were pumped out through a pipe leading from the building into neighbors' dwellings. The
ACE treatment design is to clean up all surface contamination and construct a pump and treatment system for
all groundwater contamination. After successful completion of the cleanup, all materials used are to be decon-
taminated.

Water used in the decontamination process has to be
gathered and transported for treatment at a nearby
approved wastewater plant. On April 1, 1993, during
final cleanup, Eric and a few coworkers decide to cut
work short. Instead of having the contaminated water
transported to the treatment plant, they dump it in a
neighbor's backyard. Unbeknownst to them, the owner
of that property, Rob, witnesses the entire act and takes
pictures of the parties involved in the dirty deed.

Upon hearing from Rob, ACE promises him that they
will take care of the matter and that these individuals
will be fired. After weeks of no progress, Rob informs the
Department of Environmental Protection. The DEP contacts ACE about the matter and informs them that if corrective measures do not commence, legal action will be
taken.

Questions for Discussion:

1. What are the facts in this case?

2. What health hazards do solvents cause?

3. What actions could ACE have taken to ensure that this did not happen?

4. Should the DEP monitor the company's work on other projects after this incident?

5. What legal issues are involved here, and would Rob be advised to take action against ACE?

6. Health and safety plans are reviewed by the DEP or EPA, if required. Should governmental agencies have been oversight viewers of these operations?

Part *II* | *Engineering Ethics*

4

Chemical Engineering

4.1 **Case Title:** A Decision from the
Heart (rd)
Case Type: Chemical Engineering

Fact Pattern

Bill is a senior engineer in the biomedical division of a
major corporation and the head of a research department
that specializes in the construction of artificial organs.
About a year ago, an artificial heart was created and
tested in a human patient; unfortunately, the patient sur-
vived for only nine months. However, it was a huge step
in the fight against heart disease, and it brought worldwide recognition to his company and himself.

Bill is extremely proud of the progress he and his staff have made. He remembers how much work it took to convince the board of directors to support him financially in such an experimental field. The risks of failure were immense. But he and they know that a successful intervention could double the company's net profit.

One day, Mary, his top research assistant, reports to him that a problem has been detected in the tricuspid valve of the artificial heart model. With further testing, it is discovered that the rate at which this valve allows blood to pass tends to slow down after eight months of continuous usage. The coroner's report states that the patient's death was due to the body's rejection of the artificial heart. However, it is very likely that the patient's death was brought on by this flaw in the artificial heart.

Bill becomes extremely worried. If he tells his superiors this piece of information, there is a great possibility that the project will be terminated. And, if this knowledge becomes public, not only will it bring humiliation to the company and probably cause his dismissal, but also the company will be highly susceptible to a million-dollar lawsuit by the patient's family. If Bill decides to withhold this information, a new model could be created with the flaw corrected, without anyone knowing.

Bill decides to ask for advice from two of his dearest friends. He first asks Bob, a fellow chemical engineer and someone who understands the technical aspects of the project.

"You have no choice, Bill," replies Bob. "You made a mistake and now have to suffer the consequences. Also, if you withhold this information and it is discovered later, the situation becomes ten times worse."

Bill then phones his sister, Sheila, and explains the situation to her.

"It's a tough call, Bill," replies Sheila. "Ordinarily, I would say let the truth be known, but now that your wife's rare cancer has spread, you cannot afford to lose your medical benefits or Helen will be denied treatment."

Questions for Discussion:

1. What are the facts in this case?

2. What ethical dilemma has Bill encountered?

3. How much of a factor is Helen's medical condition?

4. Is Bill being selfish if he does not report the flaw in the artificial heart?

5. If Bill's wife did not have cancer, is the decision any easier?

6. Should Bill's loyalty to the company play a major role?

7. What do you think Bill will do?

4.2   Case Title: Faster Is Not Always
Better (rd)
Case Type: Chemical Engineering

Fact Pattern

Maria is a process designer, twenty-nine years old and seven years removed from college. Like most ambitious chemical engineers, she works hard and hopes that such dedication will be noticed and result in a promotion. Unfortunately, that opportunity has never presented itself; she becomes frustrated and decides that a change of scenery is in order.

Maria applies for a managerial position at an up-and-coming engineering firm. She is hired, but her salary is actually lower than at her previous place of employment. However, through stock options, she is compensated with 2 percent ownership of the company.

In her new job, Maria oversees the development of a plastic that is becoming more popular in the field of construction. It is lighter, more durable, stronger, and less expensive to manufacture than what is currently being used in the industry. The chemical reaction that produces this plastic requires a certain catalyst in making the plastic so strong; however, the reaction converts only 60 percent of the reactants.

One day, Maria's supervisor, Tom, proposes a change in the manner in which the reaction is run.

"Listen, Maria," Tom says, "the company is looking for new ways to increase profits and gain a larger percentage of our product's market. I have an idea, but I do not know if it's such a good one. Maria, what if we replaced the current catalyst with a cheaper one?"

"That would be great," Maria states.

"However," Tom replies, "it would produce a product that is only 85 percent as strong as the original."

"Then we cannot do it," Maria comments. "Even if the product was only 5 percent weaker than the original, the risks would be too great."

"But the deficiency can be detected only under extreme tests. And, this new catalyst is cheaper, causes the reaction to occur twice as fast, and has a conversion factor of 80 percent. Also, by producing more plastic at a faster rate for half the cost, the company could sell the product at a price 20 percent lower than the competition's price. As a result, there would be a greater interest in the company on the stock market, thus improving our financial situation," answers Tom.

After some thought, Maria decides to go along with the plan. Two years later, approximately a year after Tom has retired, she discovers that several buildings constructed by one of the companies that purchased their product have been destroyed by a severe earthquake. The next day, Maria is asked by her new supervisor if she will testify on behalf of her company in a civil suit.

Questions for Discussion:

1. What are the facts in this case?

2. What ethical dilemma has Maria encountered?

3. Are Maria's intentions selfish or are they for the benefit of the company?

4. Is Maria as guilty as, less guilty than, or more guilty than Tom?

5. If a construction company was aware of the strength deficiency, would it be ethical for Maria to sell them the product knowing that people's lives could be endangered?

4.3   Case Title: Old Secrets in a New
Job (se)
Case Type: Chemical Engineering

Fact Pattern

Tim, a chemical engineer with four years' experience in offset printing processes, has been hired recently as an engineering supervisor in the printing product division of XYZ company. Until now he was employed as a research chemist by a competing firm, and during the past two years he personally developed a new formula and manufacturing process for press blankets. The new blanket is now available and gaining an increasing share of the market for Tim's former employer.

In the offset process, the rubber blanket cylinder on the press receives the image from the inked printing plate and transfers this image to the paper. The blanket is thus an important determinant of printing quality. Tim's formula and manufacturing process resulted in a blanket that not only produces superior-quality images but also wears longer, reducing the cost of materials and the cost of press downtime for blanket changes.

XYZ executives who interviewed Tim for his new job made no mention to him of the new offset blanket. They indicated it was his managerial potential that interested them, since the company would be expanding and soon would need many more managers with scientific experience than were available at that time. Tim was anxious to move out of the laboratory and into management work, but his former employer did not afford him the opportunity.

The responsibilities of supervision and administration have brought Tim to grips with new kinds of problems, as he hoped would be the case. However, one problem, currently sitting on his desk in the form of a memo from the division's director of engineering, is causing him particular concern. It read as follows: "Please see me this afternoon for the purpose of discussing formulas and manufacturing processes for offset press blankets."

This is the first reference anyone has made to the use in his new job of specific technical information from his past. Tim realizes he will have to decide immediately to what extent he will reveal data concerning the secret processes being used by his former employer.

Questions for Discussion:

1. What are the facts in this case?

2. What issues are involved?

3. What would you do if you were Tim?

4. What is the ethical issue?

4.4   Case Title: Mary's Car Trouble (jl)
Case Type: Chemical Engineering

Fact Pattern

Mary, a junior chemical engineering student, owns a late-model car, which has come due for inspection. According to a recent state law, all cars must now pass an emissions test in order to pass inspection. Ray, a mechanic, performs the test.

"Mary, I have some bad news," he says.

"What?"

"You didn't pass the emissions portion of the inspection. I'm pretty sure it's the catalytic converter. Unfortunately, it will cost you five hundred dollars to get it fixed."

"I really don't have that kind of money to invest in this car," Mary protests.

"Well, you weren't that far off from passing the test. Tell you what I'm going to do. If you give me twenty dollars, I'll forget about the test and pass the car. Who will know? Besides, it's not like you're spitting out blue smoke or anything."

Questions for Discussion:

1. What are the facts in this case?

2. What is the ethical problem facing Mary?

3. Do you think Mary should give him the money?

4. If the environmental pollution was more severe, would it affect the morality of the situation?

4.5   Case Title: Under Pressure (jl)
Case Type: Chemical Engineering

Fact Pattern

Michael's company is currently in the process of designing a chemical plant. Michael has been given the job of designing the emergency pressure relief system for one of the plant's reactors, which operates at high pressure as well as high temperature in order to achieve a high single-pass conversion. The design requires that two high-pressure valves be used to vent the gases in the reactor should the pressure exceed the upper design limits.

The engineering company contracts UP, a company that markets single valves. The valve company has had some problems in the past with their line of high-pressure valves, but they assure Michael that their valves have been tested and passed. Michael realizes that if a reaction were to proceed uncontrolled and if the pressure relief valves did not function properly, the result would be disastrous. But redesigning the system to use several lower-pressure valves would push back the completion date of the plant as well as cost the company more in terms of capital and maintenance costs.

Michael decides to use the high-pressure system. A week after the plant is started, the reactor pressure exceeds the upper limits. The valves fail to open, and the resulting explosion kills a man. After an official investigation, it is determined that the explosion was due to operator error, and no company benefits will be paid to the victim's family.

Questions for Discussion:

1. What are the facts in this case?

2. What is the ethical problem facing Michael?

3. Do you think Michael or the valve company is to blame?

4. If there was no accident, would the morality of the decision change?

> **4.6   Case Title:** Don't Judge a Chemical by
> Its Label (mr)
> **Case Type:** Chemical Engineering

Fact Pattern

Steve is an employee of a chemical company who is working in the company's exportation office. He is responsible for making sure that the labels of the products they export to Canada agree with the material safety data sheets on record at the company. If information is missing from the product labels or incorrect, some substances illegal in Canada could be inadvertently admitted into the country.

In doing some interoffice paperwork, Olivia, who works in the office where Steve is a supervisor, discovers a discrepancy between the safety data sheets held by their company and those used for the labels of a shipment of chemicals sent out to a small pilot plant in Vancouver last week. Olivia mentions the discrepancy to Steve, who signed his approval to the shipment with the wrong labels. On top of that, the customer in Canada called a few days before shipment to check with Steve that the products were approved for importation; they had a lengthy discussion, since the customer was not a chemical expert but rather a design engineer involved with the start-up of the Vancouver plant. The discussion, Steve realizes, was based on incorrect safety data.

Since the Canadian customers seem unaware of the discrepancy, Steve would like to keep the mistake quiet and not risk his job. In order to cover his tracks, he asks Olivia not to worry about it. Since she seems reluctant to ignore it, Steve promises to give Olivia a recommendation for a departmental promotion that he knows she's planning to apply for. If she doesn't help him cover up the mistake, Steve says, he can't help *her* out.

Questions for Discussion:

1.   What are the facts in this case?

2.   How is the discrepancy in the material safety data harmful to those involved?

3.   What is Steve's dilemma in dealing with the problem?

4.   What choices does Olivia have in responding to Steve's suggestion? What would be the ramifications in each case?

4.7   Case Title: Who's to Blame? (js)
Case Type: Chemical Engineering

Fact Pattern

A chemical plant on the West Coast has had a serious reactor explosion that resulted in several injuries and a fatality. There is an ongoing investigation to find out what caused the explosion. Injured workers and their families are angry and anxious to blame someone for this unfortunate accident.

Stan, one of the head executives at this chemical plant, is trying to get to the bottom of the situation with Terry, a plant manager.

Stan asks, "So, did you find out anything new?"

Terry responds, "Well, the reactor that exploded was recently upgraded to increase production. Everything was supposedly checked out, but someone overlooked a minor factor. The reactor eventually overloaded and exploded."

Stan asks, "Well, who's responsible for that mistake?"

Terry disappointingly says, "Well, remember Laura? She was one of the chemical engineers working on that production line. She quit about a month ago. Everyone seems to be pointing fingers at her, saying that she was in charge of upgrading that reactor."

"I have all of these angry people ready to seek revenge for this explosion. Terry, the reputation of this company is at stake. Not only have we caused injuries, but we have polluted the environment. Now you tell me the person at fault isn't even here anymore. How is that supposed to help anything?" shouts Stan.

Reluctantly, Terry suggests, "Well, we could always blame someone who is dispensable. For example, Rob, one of the janitors that works here, has been acting up lately. His supervisors have warned him about slacking off. He has quite a temper and tends to scream threats at people. We could try to blame him."

Stan asks, "But how would we do that? Wouldn't people know that it's not his fault? Can we get away with that?"

"Well, most people wouldn't be surprised Rob got mad about something and decided to mess up one of the reactors. He was working the night the explosion occurred. In fact, the explosion happened during one of his shifts. Maybe he was at the other side of the plant when it occurred, which is probably why he was not hurt," says Terry.

"But isn't this blatantly wrong? We cannot blame an innocent man for this," Stan comments.

"Stan, we are in a serious mess here. Our company's reputation is on the line, you said so yourself. We may all lose our jobs. No one will care if we blame a rowdy janitor for this. He was bound to be fired anyway. His attitude was not going to be tolerated much longer. I think that this is the best solution for everyone," says Terry.

After pondering the situation for a few minutes, Stan agrees. "After weighing all of the pros and cons, I think you're right, Terry. Our blaming Rob for the explosion may not be the moral thing to do, but we have to try to solve this problem by hurting as few people as possible. If we do it this way, we might be able to recover from this disaster."

Questions for Discussion:

1. What are the facts in this case?

2. What other options do Terry and Stan have to remedy the situation?

3. Was the decision to blame Rob justified?

4. Do you think Terry and Stan can get away with blaming Rob for the explosion?

5. If the truth was revealed, what do you think would happen to Stan, Terry, Rob, and the rest of the company?

4.8   Case Title: Don't Get Excited
over Nothing (js)
Case Type: Chemical Engineering

Fact Pattern

Bob is an engineer managing one of the most efficient
production lines in a chemical plant. Recently, Bob's
supervisor, Mr. Jay, complimented Bob on the great work
he has been doing. Mr. Jay even assigned Bob the task of
upgrading the equipment on the production line. More
than happy to take on the assignment, Bob begins to put
a lot of time and effort into his new project, researching
the new equipment and technology available and preparing a budget. He is working hard because a promotion might be his reward if all goes well.

One day, Bob happens to reads a report on a particular
chemical that has just been determined to be carcinogenic. The report states that this chemical could be hazardous to humans if they are exposed to it in
concentrated amounts or for extended periods of time.
Bob knows that the chemical is a minor by-product on
this production line he is upgrading. He is a bit concerned about the health risk, but he knows that the chemical is produced in small, unconcentrated quantities. He
also knows that this chemical is then used in another production line and converted to a harmless substance.

Although there is a health risk, Bob considers it insignificant. In fact, he thinks it's not even worth mention-
ing. If anyone has a concern, he would gladly assure
them that the plant is safe. It meets all regulations pertinent to the chemical. Besides, it would cost money and
time to redesign the production line, which is currently
profitable. After all of the hard work Bob has put into

this project, it would be a shame to shut down the line for such a negligible risk.

So, Bob does not mention the chemical report to anyone and goes ahead with the upgrading. He successfully executes the project, which results in an increase in production. Impressed by Bob's hard work and dedication, Mr. Jay rewards him with a promotion.

Terry, Bob's coworker and friend, goes to congratulate Bob on his promotion. While talking to him, she mentions the chemical report.

Terry says, "I've been meaning to talk to you about this report. Doesn't your production line make this chemical as a by-product? Aren't you concerned about the health risk?"

Bob replies, "It's only a minor by-product. The concentrations are too low to provide any serious risk. Don't worry. And please don't mention the report to anyone else. I do not want people to get excited over nothing."

Questions for Discussion:

1. What are the facts in this case?

2. Do you think there is a considerable health threat involved in this production process?

3. Should Bob have kept the report a secret?

4. Should Bob have consulted anyone before making his decision?

5. How much should the issues of money and effort have weighed in Bob's decision?

6. Does Bob deserve the promotion?

4.9   Case Title: It's Not My Concern
Anymore (js)
Case Type: Chemical Engineering

Fact Pattern

Laura is an engineer working in a chemical plant. She has recently received a job offer from another company, which she accepts because she knows that the new job could be a big step in her career.

Laura is responsible for one of the production lines in the plant she will soon be leaving. She has always been a reliable worker and an effective manager. However, having handed in her letter of resignation, she has been less attentive to her work over the past couple of weeks. She figures that there is no need to worry about this job anymore; she has to concentrate on her future.

On Laura's next-to-last day of work at the plant, Harry, a coworker on the same production line, finds out that there is a problem with the purity of the product: The level of impurities is a little higher than acceptable. Harry decides to consult Laura.

He says, "The product coming out is below the required purity. I think you should investigate it so we can solve this problem."

Laura replies, "I would love to help you, Harry, but tomorrow is my last day here. I don't want to start dealing with this problem; it could take a while to solve. Let my replacement worry about it."

Harry answers, "Laura, if we let this problem go, we'll continue to have a product that doesn't meet regulation. The problem could also get *worse*. You are the expert here, so you could easily fix this mess."

"Harry, you're a friend of mine. Please don't ask me to get involved in this problem; it's not my concern anymore. I just want to relax during my last two days at work," pleads Laura. "It's not like the plant will blow up. Wait for two days. You can pretend that you didn't notice anything until then."

Reluctantly, Harry agrees. "I know you're really looking forward to your new job. It's just that I'll feel guilty knowing that something is wrong, and I'm not doing anything about it. But I guess I can wait for two days."

"Harry, don't worry. Take it easy for a couple of days. Just think of it as a minor delay," replies Laura.

Questions for Discussion:

1. What are the facts in this case?

2. Do you think Laura should stay focused on her current job?

3. Should Laura handle the problem?

4. Do you think it's okay for Harry to ignore the problem for the next two days?

5. Should Harry consult someone else now that Laura has refused to deal with the problem?

> ### 4.10   Case Title: She's My Friend, but
> ### I Can't Tell Her (js)
> ### Case Type: Chemical Engineering

Fact Pattern

Steve works in a chemical plant and lives in a town a few miles away. The town has always had a good relationship with the plant because many of the townspeople are employees.

Recently, Steve found out that his friend, Judy, has cancer. Steve was shocked that she, a thirty-five-year-old woman, developed cancer at such a young age.

Judy has seen her doctor, who has noticed a recent occurrence of cancer developing in middle-aged people in the town. In an attempt to find a reason for the new cancer trend, the doctor has asked her about daily patterns, food intake, and places she goes; Judy has provided the necessary information. Her doctor has not been able to find a common pattern among the new cancer patients that would explain the town's cancer "hot spot." Judy decides to talk to Steve about her situation.

She says, "My doctor could not see anything unusual in my daily life that might have caused the cancer. I haven't been extensively exposed to anything carcinogenic."

Steve responds, "Maybe it just happened. Just because a few people in an area develop cancer around the same time, it does not necessarily mean that there is something in common that caused it."

The next day at work, Steve remembers that the plant is doing some testing on a new chemical, which might be carcinogenic. Steve thinks that there might be a leakage or emission from this chemical testing that is slowly polluting the adjacent area, which is full of stores and restaurants that mostly middle-aged people go to. Although Steve has no evidence to back up his belief, he begins to think that this might be the cause of the recent cancer trend.

Steve wants to tell Judy about his hunch. However, he is afraid to squeal on his company and lose his job, especially since he has no real evidence to support him. Also, if his hunch is incorrect, the close relationship between the township and the plant could be unnecessarily tarnished. The people of the town would be alarmed to find out that the plant is not being careful about leakage or emission.

Although Judy is a friend, Steve cannot afford the consequences of telling her his suspicion. So, Steve decides not to. Besides, Steve believes that if there is a leakage from the plant, it will eventually be noticed. In fact, Judy and her doctor might actually figure it out with a little more investigation.

Steve decides to help Judy come to the same conclusion by dropping a few hints. He probably will do a little more investigating at the plant to make sure everything is as it should be. This way, Steve believes that he can keep his job, help his friend indirectly, and get enough evidence to support or put an end to his suspicion.

Questions for Discussion:

1. What are the facts in this case?

2. Should Steve tell Judy about his hunch?

3. Do you think that Steve is more concerned about himself or the town?

4. What is your opinion on Steve's decision to subtly help Judy and to do a little more investigating?

> **4.11   Case Title:** Poor Quality
> Makes a Sale (gt)
> **Case Type:** Chemical Engineering

Fact Pattern

Bill is a sales representative for an equipment replacement part company. Since he works on commission, he makes calls to potential customers to inform them of sale promotions. However, Bill does not inform his customers of the possible dangers in using these "on sale" replacement parts. The parts he is selling are components of cheap, poor quality, with higher failure ratings than good-quality parts.

His assistant, Jeff, takes the orders from Bill's customers and follows up on them. He makes sure that the components are sent to and installed at the customer's site. One day, he finds that some of the replacement parts are defective and informs Bill of this.

"Bill, I found a couple of gauges to be defective in one of the ordered lots. Should we return them to the manufacturer?" asks Jeff.

Bill responds, "Just throw out the defective ones, and we'll ship those remaining to our customers. They won't know about it."

"But what if other gauges have defects I can't see? The component may fail and cause an accident," replies Jeff.

"That won't be our problem. Besides, there hasn't been an accident yet. Don't worry about it," says Bill.

Jeff is nervous about the issue, but Bill is his boss. Should Jeff ignore it and continue to do his job? What if an accident does arise? Should Jeff confront his boss and risk his job?

Questions for Discussion:

1. What are the facts in this case?

2. What should Jeff's next action be?

3. Is there a solution to Jeff's dilemma that he and his boss can agree upon?

4. Who will be responsible if an accident *does* occur?

> **4.12   Case Title:** What's Going On? (gt)
>   **Case Type:** Chemical Engineering

Fact Pattern

Tom works in an engineering company and is responsible for managing the production of its petroleum product. His boss, Ken, informs him of a need to increase output of that product because customers have been placing more orders than usual.

Tom continues to socialize with his coworkers as usual. He thinks that speeding production is just a matter of cutting some procedures short; Tom decides to eliminate the extra cleaning step in the process to obtain a quicker turnaround time.

The company later receives customer complaints regarding the quality of the petroleum product; it was contaminated with cleaning chemicals that were supposed to be removed prior to shipment. Ken meets with Tom to discuss the issue.

"Tom, we've been getting customer complaints regarding our product," says Ken. "They have found the petroleum contaminated with other chemicals. What's going on?"

Tom is left confused on how to get out of this predicament. What can he say to Ken without risking his job and losing his production team? Should he blame his team? Or should Tom tell Ken the truth?

Questions for Discussion:

1. What are the facts in this case?
2. How should Tom respond to Ken's question?
3. How could Tom have avoided his problem?

4.13 Case Title: Where's the Quality
in the Quality Control
Group? (gt)
Case Type: Chemical Engineering

Fact Pattern

Jane works in the quality control group in the engineering section of her company. She is responsible for testing samples in her group's backlog. Much of the backlog contains similar samples that all need to have the same testing. This testing procedure takes hours to conduct for a single sample.

Jane has worked for three weeks straight on the same batch of samples; all of the tested samples have met quality control standards. With a holiday weekend approaching, she decides to end her week early by entering similar results for untested samples.

Glen also works in the quality control group with Jane. He sees her forging numbers in the computer and pulls her to the side.

"Why are you putting in results when you haven't finished testing the remaining samples?" inquires Glen.

"I've got a big weekend planned, and the samples in this batch seem to be routinely passing the tests," Jane replies. "The rest of the batch is probably going to pass testing anyway. I'm just saving myself some time to prepare for the weekend."

"What if the rest of the samples really don't pass the testing?" asks Glen.

"I test these types of samples all the time," responds Jane. "They always pass quality control testing."

Glen is troubled with the issue and wonders whether he should just leave it up to Jane.

Questions for Discussion:

1. What are the facts in this case?

2. What should Glen do?

3. Are there any other options for Jane? For Glen?

> **4.14   Case Title:** School Lunch Safety (gv)
> **Case Type:** Chemical Engineering

Fact Pattern

Anne has been hired at a leading food distribution company that caters to elementary schools around the nation. The company feels they have been spending too much money on packaging, which includes foil-lined paper, so that the food may later be heated in the package.

Anne and her supervisor, Tom, hire a group of consulting engineers to research more economical ways to package the food that would still allow it to be heated. The engineers research as many packaging alternatives as possible and come to a conclusion. They explain to Tom and Anne that there are a variety of packaging options available, but all of them are just as expensive as the one being used. However, the engineers explain that if they were to use plastic packaging, the company could save a substantial amount of money. Tom is very pleased and asks Anne to switch the packaging as soon as possible.

After the meeting Anne says to the engineers, "Thanks for your help. The money you helped us save means we can hire more workers and also earn more money in the future."

Anthony, one of the consulting engineers, says to Anne, "There is something you should know. Although the new packaging method is cheaper, it contains a polymer called polyvinyl chloride that has been linked to cancer and can be very toxic. When the package is heated, the PVC most likely will be released into the food."

Anne replies, "So, you are saying we are contaminating the people who eat the food without telling them."

"Yes, but a majority of the companies do it and no one has complained yet. If you do *not* use this new packaging, your company will lose a lot of money, and you may ultimately lose your job."

Questions for Discussion:

1. What are the facts in this case?

2. What issues are involved?

3. What are the choices that Anne must make?

4. What could be the consequences of each decision she could make?

> **4.15 Case Title:** HairMagic Company (gv)
> **Case Type:** Chemical Engineering

Fact Pattern

Joe Murray, a chemical engineer at the HairMagic Company, is in charge of the production of a new hair coloring product, which is expected to be ready for mass production in a month. Joe has been reviewing the results of the hair coloring tests conducted on volunteers, which show that when the product was used on dark-haired men and women, it changed the color of the hair with an accuracy of 99.2 percent. However, on fair- and light-colored hair types, it had an accuracy of only 89.3 percent. Joe realizes that the product needs to be refined further and that some substitutions should be made in its chemical makeup.

Joe decides to go to his manager and explain that production will have to be delayed. His manager, on the other hand, feels that production should be initiated and that further refining can be done during production so as not waste any more money or time.

Ethically, Joe feels strongly that this is wrong: The refining needs special attention, and production should not even be considered when the product is not yet perfected. Joe's only alternative is to go above his manager and try to convince upper management to delay production. However, doing this could anger his manager and could also lead to Joe's dismissal. But it might also make him appear to be confident and conscientious, and could also serve to publicize his abilities as an engineer.

Questions for Discussion:

1. What are the facts in this case?

2. What issues are involved?

3. What are the choices Joe must make?

4. What could be the consequences of each decision he could make?

5

Civil Engineering

5.1   Case Title: A Deadline with No
Pressure (rtc)
Case Type: Civil Engineering

Fact Pattern

A nearby township has recently purchased one thousand
acres of land zoned for residential use. Their plans for
the near and distant future include the construction of
hundreds of houses as well as some townhouse com-
plexes, but there is no stipulation allowing for commercial or industrial construction.

Your firm has been contracted to design the aqueduct
and water distribution system for this new section of
town. Your boss has named you the project leader and

has given you a deadline, one that will require you and your team to work evenings as well as weekends to meet it.

The day before the deadline, you are reviewing the calculations one last time and come across an error. The design specifications require that each house must have a minimum pressure of 40 pounds per square inch (psi) going into it from the water distribution system; at the outermost regions of the newly acquired land, pressure delivered to each house will be less than that. To recalculate and redesign for all of the specifications would take days, and you have reassured your boss that you will meet the deadline. The boss has already scheduled a meeting with the clients to present the designs.

The pressure being delivered to these houses is only a few psi below the minimum for the design requirements, and the chances that someone will discover the error are very slim. Also, these houses will be the last constructed, in the very distant future.

The hard work and extra hours you have put into the project to meet the deadline will surely put you high on the list for the next promotion. However, if you fail to meet the deadline, all of the hard work and extra hours will have been for naught.

Questions for Discussion:

1. What are the facts in this case?

2. What is the *ethical* dilemma you are faced with?

3. If the pressure is only marginally below what the client specified, is that considered marginal error and permissible?

4. Do you have a greater responsibility to your client or to your company?

5. In either case, what course of action do you take?

6. What are your motives behind your course of action?

7. Is there a solution to the problem that would satisfy both the design specifications and your boss's demands for meeting the deadline?

5.2   Case Title: Government
Spending (rtc)
Case Type: Civil Engineering

Fact Pattern

Your firm has been contracted by the government to work on a public utility project. Specifically, the firm must complete the first part of the project, which includes determining components of construction to meet design parameters. The results of the firm's work will be turned over to various governmental agencies.

The firm happens to be going through some financial troubles at this time, and they have been overcharging the government in order to help keep the firm going. There are several smaller projects being financed by the overcharges; they are being billed as work related to the government project. So, in addition to being over-charged, the government is being billed for services not rendered.

Your firm has provided good results based on consci-entious work; they are therefore meeting their contrac-tual obligations. The firm simply is padding the budget.

There is minimal risk to your firm of getting caught because they have worked under this kind of arrange-ment with the government several times before and are highly trusted. Compared to the total cost of the project, the extra charges are rather small. However, if your firm does get caught, the professional reputations of several of your friends and colleagues could be ruined; they could be fined and maybe even imprisoned. Finally, if the firm does get caught, chances are it will be forced to fold, and you will be unemployed.

Questions for Discussion:

1. What are the facts in this case?

2. What is your responsibility in the matter?

3. Is it allowable because "everybody rips off the government"?

4. What course of action would you take?

5. If you choose to say nothing, then are you equally as guilty as those directly responsible?

6. Would your decision change if your wife/husband worked for the same firm?

7. What could be the repercussions of your decision, whatever it may be?

5.3   Case Title: Not up to Standard (brh)
Case Type: Civil Engineering

Fact Pattern

Richie has been a licensed electrician for fifteen years and was recently hired as foreman for a large New York–based electrical contractor. The company that hired him has a reputation for doing jobs at extremely low prices and for paying their workers very well.

The first day Richie works with seven other electricians who report to him, starting a remodeling project in a commercial building for a large retail chain. The supplies for the job were ordered and delivered to the job site before Richie started his new job. As Richie checks on the supplies to make sure that everything ordered was also delivered, he notices several items not allowed in commercial construction, including several rolls of cable that should be plenum rated but are not. The cable they do have costs approximately one-half the price of plenum-rated cable. At first Richie thinks that there has been a mistake in the order, so he decides to call his boss, Dave, and double-check it with him.

Dave tells Richie that even though the National Electric Code (NEC) mandates that plenum cable be used, they will use non-plenum-rated cable. When Richie asks why, Dave responds by saying that he feels the NEC requirements are too strict and that cost of the cable is outrageous. Dave also says that nobody will realize they are using the wrong cable, so it doesn't matter. Richie disagrees with Dave but chooses not to share his displeasure with his new employer.

After Richie finishes talking to Dave, he goes to work using the supplies that he has on the job site. As the day progresses, Richie notices that the other electricians, all of whom have worked for Dave for years, are taking many shortcuts. When Richie asks them about their work practices and why they are taking so many short-cuts, they respond by telling him that that's how Dave wants them to do things because it is less expensive than if they're done properly. Besides, they say, no one will know.

Richie is beginning to see how Dave can afford to pay his employees so well while keeping jobs at low cost. Richie isn't sure that he agrees with this method of doing things, but he does like his new salary and benefits. Now he is in the position of trying to decide what he should do—look the other way and ignore the violations, or quit his new job and report Dave?

Questions for Discussion:

1. What are the facts in this case?

2. What are the ethical issues?

3. What do you think Richie's decision should be?

> **5.4   Case Title:** Underestimating the
> Competition (brh)
> **Case Type:** Civil Engineering

Fact Pattern

Tom works as an estimator for a large New York electrical contractor. For the past two months he has been working on a bid for a project in upstate New York. The bid is valued, according to his estimate, at $7.5 million.

After Tom submits his bid and the bid results are tabulated and distributed to the bidders, Tom learns that his company wasn't the low bidder on the job. Another company, Windsor Electric, has bid the job for the lower estimate of $6.7 million. Windsor, after learning of the discrepancy between their bid and the bid completed by Tom's firm, decides to not accept the job, fearing that they have made a major mistake in the bid proposal.

After Windsor Electric withdraws its bid, the owner of the project decides to allow both Windsor Electric and Tom's company to rebid for the job. The next day Tom is in the office talking with his boss, Jeff, and another estimator, Dave, about the best way to work on the project to make sure that they get the contract. They look at several ways of installing and using different materials, but to no avail. Their estimate would still be well over seven million dollars, no matter what they do. It is at this point that Dave suggests they sit down and talk to the president of Windsor Electric in the hope they can work out a mutually agreeable way of rebidding the job. What Jeff really wants to do is pay Windsor Electric one hundred thousand dollars to raise their bid to a higher price than Tom's company, which would allow Tom's company to be awarded the contract.

Tom isn't sure if this is an ethical approach to use in order to be awarded a contract, but both Jeff and Dave feel that since the arrangement is beneficial to both companies, there is nothing wrong with doing it.

Since Tom is the only person in the office who is knowledgeable about the project, he will have to be the one to speak with the people at Windsor Electric. Jeff tells Tom that if he really feels uncomfortable with the idea of paying Windsor, he will not force Tom to talk to them.

Questions for Discussion:

1. What are the facts in this case?

2. What are the ethical issues?

3. Is Jeff's argument about the benefits to both parties logical?

4. What do you think Tom's decision should be?

5.5   Case Title: Dangerous Deception
(jmsl)
Case Type: Civil Engineering

Fact Pattern

Mike has been designing buildings for almost ten years now, after going through all of the necessary education, undergraduate and master's, and working his way up the ladder in a consulting firm. Now he is the project leader for the design of a large building in the city. There are many people under him, and he is calling the shots. For the most part he is very happy with his job. All his life he has wanted to design buildings and use his skills to make a difference in the world.

Unfortunately, however, the company he is working for is rather greedy. There are five partners at the top of the consulting firm, each enjoying the high-profile, extravagant type of lifestyle that goes along with being leaders of a large, prestigious consulting firm.

The greediness also trickles down the chain of command within the consulting firm; at every chance employees try to figure out a way to cut corners.

"How is that building project of yours coming along?" asks Kevin, one of the five partners.

"Good, we're on schedule, and members of the project team are working extremely well together," replies Mike.

"There are a couple of things I wanted to make you aware of, I mean as far as the design of the building is concerned," states Kevin.

"Go ahead, I'm all ears," answers Mike.

"Well, to start with, we have this deal going with the contracting company we will be using for the construction of the building. If we specify materials of slightly lower quality, they will give us a kickback. You know what this means. Since you are the project manager, I can assure you that you will see a piece of this," explains Kevin.

"How much lower-quality will the materials be?" asks Mike.

"Don't worry about that; I'll let you know what the specifications have to be. I just want to be sure that you are OK with this," states Kevin.

"Well, I guess I'm OK, but what about the city building codes we have to meet? There's no way that they'll even let us construct this building if we are using inferior materials," says Mike.

"Don't worry about that either. I've got it all taken care of," explains Kevin.

Mike thinks this over for several days. He is very uncertain about what he should do. It doesn't sound like what they are doing is too bad, but what if injuries occur as a result of the poor-quality materials? He likes his job and really can't afford to be unemployed right now; there are just so many bills to pay.

Questions for Discussion:

1. What are the facts in this case?

2. What are some of the problems facing Mike?

3. Is there any way that he can resolve the problem without putting his job in jeopardy?

4. What type of problems can arise from using materials of poor quality?

5. What else do you think Kevin means when he says he's got it all taken care of?

5.6   Case Title: School Spirit or
Self-Interest? (cm)
Case Type: Civil Engineering

Fact Pattern

George is a student of civil engineering at a local college. He is a good student: He does his work on time, participates in class, and tries to understand all of his course material on the subject. Even though he knows his résumé looks attractive because of his academic record, he also knows that companies look at more than grade point average. He has to prove that he is a well-rounded person by committing to school activities, either curricular or noncurricular.

After researching the types of organizations that exist on campus, he realizes that he has many options, and he tries to belong to as many associations as he can. Some organizations are running well under their current leaders, but some are struggling to complete their programs.

The problems of these student groups are due to lack of members' motivation, lack of support from faculty members, and an overall lack of organizational purpose. George sees these problems, but he does not try to help to solve them. As long as his résumé records his membership in these organizations, he does not care.

At the end of the school year, there is a need for new leadership in the school organizations. The graduating student leaders try to get some people to take the executive board positions in their respective organizations. One of the school rules is that one person cannot hold office in two or more organizations at any given time. Even though the rule is usually followed, some students *do* take more than one student leadership position. Since the faculty and staff of the college do not want to bother with the amount of paperwork involved in the removal of a student from office, students get away with leading more than one organization.

George wants to take advantage of this situation. He will be taking twenty-two credits the following year, so his school schedule will be very busy. Despite this workload, he decides to run for many offices, in more than two school organizations.

He knows that the extra organizations will be a great asset on his résumé, and he hopes to win at least two leadership positions. Since very few people even try to run for executive board positions, George is elected for various positions in three organizations.

George knows that he will not be able to handle school *and* all these extra activities. He is asked by a faculty member to decline some office positions, but he wants to retain these positions on his résumé. The teacher tries to convince George, but after a while, he gives up.

George decides not to step down from any position. He feels guilty, but he knows that if he wants to get a job when he graduates, he needs those leadership positions on his résumé, even though in practice he won't live up to the responsibilities therein.

The following year, many student organizations are run by a very small group of people. If this situation continues, the organizations will become less diverse year by year.

Questions for Discussion:

1. What are the facts in this case?

2. What is the ethical problem that George is facing?

3. What are the ethical problems that the teacher faces?

4. Do you think that the school staff should be more active in school activities?

5. Do you think that industry's search for student leaders should affect George's decisions?

5.7   Case Title: I Know It Anyway (gr)
Case Type: Civil Engineering

Fact Pattern

Bill is a new hire at a chemical engineering lab. He has been told by his superiors and by the fire department that in order to supervise the lab, he must get a certificate of fitness from the city. Bill has studied the pamphlet that the city gives out and makes an appointment to take the test. He is not nervous at all about this test, which covers some basic chemistry, some memorization of building codes, and plain old common sense.

It takes him a while to get down to the building where the testing is done, and so far it has proven to be a lousy day. Bill waits in all of the necessary lines for a long time, which is typical, and finally gets to sit down in a big room and take his test. He flies through the first half of it before he notices that most of the answers are already marked off in pencil, and they happen to be correct! Bill knows that he has done most of the test already, and he is very confident that he knows all of the answers. He thinks about returning the test book and requesting a blank one, but he has already been downtown too long, and that hassle will surely keep him here for a few more hours. Bill just covers the marks with his hand and continues on through the test. It is scored in front of him via machine; he gets a perfect score and is handed his certificate.

Questions for Discussion:

1. What are the facts in this case?

2. What are the ethical issues in this case?

3. Does Bill's knowledge of the answers independent of the marked-up test copy mean that this is not cheating?

4. Should Bill have acted any differently?

5. What would you have done?

5.8 Case Title: Stormy Weather (pw/cc)
Case Type: Civil Engineering

Fact Pattern

Hank is a young civil engineer fresh out of school and working for a major design company in Manhattan. He has finished the design for his first project with the firm, an addition for an elementary school located outside the city, and is happy that everything went relatively smoothly. Since it was his first experience working on a real project with real deadlines, he often felt rushed and was unable to put the amount of time into the project that he would have liked. He feels somewhat reassured, however, by the fact that his design was checked by his supervisor, even if it was only a quick glance at his numbers.

Hank has some time on his hands before his next assignment is to begin, so he decides to go over some of the drawings from his first design, which was relatively simple but gave him some valuable experience. After examining the drawings, Hank realizes he made a mistake in his design. He used the wrong building code for his design and therefore used the wrong allowable wind design. Instead of designing for 120-mile-per-hour winds, he designed for only 75 mph. Hank is dumbfounded that both he and his supervisor could have missed such a detail.

The construction of the project is almost completed. Hank knows that any adjustment made to the building now will be extremely expensive and his firm will be the liable party. He also knows that informing his boss could jeopardize his position in the firm.

What are the chances of such a strong wind coming and knocking down the building anyway?

Questions for Discussion:

1. What are the facts in this case?

2. What degree of responsibility does Hank have? His supervisor?

3. Where does one draw the line between safety and cost?

4. What do you think Hank should do?

5.9 Case Title: New Kid on the Block
(pw/cc)
Case Type: Civil Engineering

Fact Pattern

Lou is a recent college graduate who works for a local construction company. His job is to obtain a sample from any concrete used on the job and to test it for various properties. Basically, Lou is required to make sure that the concrete contractor hasn't cut any corners with the product. The job has been fairly simple so far, even boring at times, which probably explains why Lou is looking for a more challenging and better-paying position.

At his present assignment, Lou is waiting to obtain his next sample when he is approached by one of the concrete truck drivers, who offers Lou five hundred dollars if he will lie and say the concrete from his truck passes inspection. The driver tries to convince Lou that there is no way you can tell the difference between his concrete and the concrete specified by the engineer. The older man further says that this sort of thing is done all the time and is one of the big perks of being a concrete inspector. Lou asks the driver if the structure will still be safe, and the driver assures him that it will.

Lou wonders if he should take the money, figuring he can always lie and say that he tested the concrete and it passed his inspection. He knows that these structures are always overdesigned anyway, so no one would be in jeopardy. By taking the extra money, Lou feels he can make up for the small salary he is earning with his company. How else is he going to pay off his college loans?

Questions for Discussion:

1. What are the facts in this case?

2. What ethical problem is Lou facing?

3. Do you think Lou's financial situation justifies making a decision in favor of corruption?

4. Is the truck driver's advice logical?

> **5.10   Case Title:** All in the Family (mb)
> **Case Type:** Civil Engineering

Fact Pattern

Henry, twenty-eight, is a senior associate in a structural design firm. His father owns a steel warehouse and does business mostly with large companies. Henry usually does most of his business transactions with his father's warehouse. Jackson, an entrepreneur, has hired the firm that Henry works for to build a five-story apartment building in Dallas, Texas. Jackson deals with Henry as the project progresses, and Henry is on call whenever a problem arises in the construction.

It is up to Henry to find a warehouse that can supply the steel for Jackson's building; he chooses his father's business. At one point in the design, the steel needed for the construction is no. 14 bars, which the warehouse does not contain; however, no. 11 bars are available. Henry thinks they can be used safely, but it is not the best design or the best price.

Henry does not know whether to use the available no. 11 steel or to buy no. 14 from another warehouse. Since he has three days to decide, he starts weighing the pros and the cons of both types.

Questions for Discussion:

1. What are the facts in this case?
2. What are the ethical issues in the case?
3. How ethical are Henry and his father?
4. What is the nature of Henry's confusion?
5. How could Henry find the best way out of that case?

6

Electrical Engineering

Fact Pattern

Sam earned his degree in electrical engineering with a high GPA a year ago. In his search for a place to work, he tried everything but couldn't find a job, so he went back to school part-time to get his master's degree in electrical engineering. Finally, he got an offer from the company Oceanic Air.

A small company with very few resources, Oceanic Air has a habit of hiring individuals without some or all of the required licenses. They offered Sam the job of an avionics engineer (aviation electronics engineer) without the required Federal Communication Commission license, which is called a general radio-telephone operator's license. Instead, the company gives Sam a period of six months to acquire the license on his own; they are too small to run deep training courses for employees.

Five months pass but Sam is so busy with working, school, and his personal life that he never has the chance to study for the FCC test, which consists of two parts: The first concerns operating the communication radio of a vessel, the second part involves basic electrical knowledge.

Jim, a coworker of Sam's, gives him an idea. He mentions that since the testing process has been privatized, he can be given the test by a private instructor for forty dollars—*but*, for an additional two hundred dollars, he will be guaranteed to pass the test.

Sam doesn't know whether he should take the test and rely on his existing knowledge, some of which is rusty, *or* study for the test and sacrifice lots of his personal and social time, *or* just pay the extra cash and relieve himself of the problem. After all, whatever information he will get out of studying for the test may just be a repeat of his college years, or it may not have much use in the real world.

Questions for Discussion:

1. What are the facts in this case?

2. What should Sam do?

3. What is the ethical question?

4. What would you do if you were in Sam's situation?

6.2   Case Title: Who's Right? (ha/nn/rv)
Case Type: Electrical Engineering

Fact Pattern

Dr. Keen, a well-known professor at MIT, conducts new research in electrical engineering. One of Dr. Keen's most prominent roles at MIT is teaching most of the graduate-level courses. Upon entering their final graduate-school year, most students must work on a single project of their own choosing or one suggested by a faculty member. Throughout the years many innovative projects have been created and, for the most part, have been supervised by Dr. Keen.

In addition to Dr. Keen's teaching, he also has a small commercial electrical engineering company. Many of the products Dr. Keen's company has produced have been similar to those the students have worked on. While most of these products could not have been developed except for Dr. Keen's assistance, some of these products are direct results of students' original ideas.

For one particular project, Keith Dell has enthusiastically worked on creating an AM radio receiver using a phase lock loop (PLL). The only problem with introducing PLLs into such devices as receivers is their cost; radio technology has typically relied on great transmitters and cheap receivers. Keith's work shows great promise. It is brought to his knowledge that Dr. Keen is going to use Keith's idea in his private business. Keith has expressed frustration over the exploitation of his labor so that Dr. Keen can make a profit; however, he somewhat believes that the learning process should be sufficient compensation. Despite Keith's mixed feelings, he *does* believe Dr. Keen should not be entitled to any rights to their projects, although prior to beginning a project, each student must sign a contract waiving all rights. Keith fears that any attempt to expose or confront Dr. Keen's behavior would jeopardize his grade.

Questions for Discussion:

1. What are the facts in this case?

2. How should Keith face the situation, and why?

3. What will be the consequences of Keith's actions?

6.3 Case Title: Intellectual Property (cc)
Case Type: Electrical Engineering

Fact Pattern

Mike has been working for a large defense company for five years and has been thinking of looking for another, higher-paying job. The department he works for makes radios for the military; he writes the software that controls Digital Signal Processing (DSP) chips in the units. Mike also works on data encryption and searches for better ways of making transmissions more secure so enemy forces cannot decode the signals.

Just recently Mike has thought of a better way of encrypting transmissions. His idea has a wide variety of applications in the telecommunications field and in cellular phones. Mike knows that according to his employment contract, any inventions he comes up with while working for the company are the intellectual property of the company. Mike would receive only a dollar from the company for his patent (due to legal formalities), and the company could stand to make millions.

Mike knows that he can leave the company in six months when his employment contract expires, and he will be able to patent his idea and market it to Bell Atlantic, Sprint, or NYNEX. But Mike knows this will be unethical and unlawful. He knows that the company may sue him, saying the invention was discovered while he was still in their employ and therefore belongs to them. Mike doesn't know what to do.

Questions for Discussion:

1. What are the facts in this case?

2. What is the ethical dilemma?

3. Does Mike deserve more than one dollar for his invention?

4. What do you think Mike should do?

6.4 Case Title: Investing, but in What? (cc)
Case Type: Electrical Engineering

Fact Pattern

Frank is a young electrical engineer. He works for Retina Scan Technologies, a small start-up firm of about twenty-five employees. The firm's main product is a piece of equipment that electronically scans the characteristics of a person's retina into a computer. An individual's retina is unique in the same way fingerprints are unique. This retina data from one person then can be matched against the data from other people's retinas.

This technology has a market with law enforcement agencies across the country. The main problem with the technology at this stage of development is the slow speed at which the database is searched. The Federal Bureau of Investigation has set a standard time of four minutes in which the retina data from one eye must be matched correctly against a database of one hundred thousand others. Retina Scan's product at this time takes seven minutes and twenty-three seconds to perform the FBI benchmark.

During the six years of research and development of its product Retina Scan hasn't had any revenues from product sales and has had to rely on investors in order to pay the basic operating costs of the business. A year and a half ago the company's funds ran low, and Frank didn't get paid for a month's work, even though he has a mortgage and must rely on a steady income. The firm has since made up this money, but Retina Scan has been looming on the brink of bankruptcy. However, Frank thinks that as long as the company stays afloat during these hard times, they will have an extremely marketable product for the future.

Every so often an investor comes to look over the facilities and talk to the engineers about the advances the company is making. One such investor is coming in next week for a demonstration of the product. Apparently this particular party has one hundred thousand dollars that he is considering investing. He wants to see how close the firm actually is to passing the FBI's benchmark. The company's CEO has asked Frank to demonstrate the product. He also explains to Frank that the company has run very low on funds again, and cannot even pay this month's electric bill. The CEO wants Frank to run the demonstration against a database of only eighty thousand other retinas instead of the one hundred thousand specified by the FBI. This will speed up the matching time to about five minutes and will show that the company has come a lot closer to meeting the FBI's standards. The CEO doesn't think that the investor will hand over any more money to the company if he knows the true current processing time.

Frank thinks that another one hundred thousand dollars would keep the company in business long enough for him to find a way to beat the four-minute processing time. He knows that faster processors will be on the market soon, and he is sure they will decrease the matching time to well below four minutes. If the FBI's specifications were met, the company would be able to sign many lucrative contracts with police departments across the country. The price of shares would skyrocket, and all the investors, including himself, would be rich.

But falsifying the test procedure troubles Frank. He has always been an honest, ethical person and normally would not take part in anything underhanded. He doesn't know what to do.

Questions for Discussion:

1. What are the facts in this case?

2. What is the ethical dilemma Frank faces?

3. Do you think that deceiving the investor could be justified?

4. If he didn't have shares in the company, would Frank act the same? Should he?

5. What might be the ramifications of going against the CEO?

6. What would you decide to do if you were in Frank's position?

6.5   Case Title: Is There a Rat in the
House? (yf)
Case Type: Electrical Engineering

Fact Pattern

"Do you think you can take a quick look at these and proof them for me?" asks Missy about the calculations she has just finished performing.

"That's no problem," responds Don, "just leave them on my desk."

Don and Missy are both employees at Manhattan House Electrical Engineering; Missy has three more years' experience (and thus a higher position) than Don. Missy has always been considered the "expert" on developing and analyzing differential equations and applying them to engineering calculations. Don too is good at doing these equations, and is the staff member to consult about them when Missy is not around.

After lunch, Don finally gets a chance to review the calculations and is quite surprised to see some of the liberties Missy took. "Making all these assumptions essentially reduces the problem to a simple, common 'textbook' situation. She really doesn't take into account all the complexities involved!" exclaims Don to no one in particular.

He begins to notice that Missy has been making fundamental calculation errors in all her assignments! In many respects, she is essentially making her calculations "fit" the actual answer. "Boy, if the boss knew about this, she would certainly be fired," thinks Don. "This discovery could mean a big advance for me, along with the respect I deserve from my fellow employees."

Don begins to turn the situation over in his head. He never really liked Missy anyway. She is pompous and talks in a condescending manner toward him all the time. But, on the other hand, isn't it wrong to turn her in without giving her a chance to explain herself? Don has to make a decision on what he wants to do.

He asks Missy about the assumptions, and she snaps at him, "That's what *real* engineers do: We make assumptions to simplify the problem. Work another three years, and maybe you'll learn how to deal in the real world."

Don is flabbergasted! Missy actually knows that what she is doing is incorrect, yet she does it anyway. He begins to wonder, "Should I just let the status quo continue, or do something about it? If I turn her in, will that engender respect from my colleagues, or will they label me as a rat?"

Questions for Discussion:

1. What are the facts in this case?

2. Should Don talk to his boss about Missy?

3. Is it unethical for Don to remain quiet when he knows Missy may be inaccurately calculating things?

4. Does Don have a responsibility, as an engineer, to disclose to the company's clients the situation?

5. Should Don just mind his own business, considering the fact the calculations Missy has been performing have worked so far?

6. What advice would you give to Don in this situation?

6.6   Case Title: Less than Full Disclosure (bg)
Case Type: Electrical Engineering

Fact Pattern

Tony Lando is a private consultant; small-business owners come to him for advice about computer needs. He
examines each company's operation and automation
needs and makes recommendations about suitable hard-
ware and software.

A private hospital is interested in upgrading the software used for patient records in the accounting department; the institution has already received proposals from
three software companies. Tony's duty is to evaluate the
proposals, looking at which system meets the hospital's
needs in terms of service, staff training, future updates,
and cost. Tony concludes that Tristar Systems offers the
best package.

He explains in his report that the software requires the
minimum amount of training for employees, provides
the best service with respect to accounting needs, and
offers a thirty-day trial-free period. Tristar also will provide a three-hour presentation of the software at no extra
cost. He mentions that the cost is slightly higher than the
others but the product is worth it.

However, Tony fails to mention that he is co-owner of
Tristar Systems.

Questions for Discussion:

1.   What are the facts in this case?

2.   Should Tony mention that he is co-owner of Tristar
Systems?

3.   What would you do?

> **6.7 Case Title:** Onboard Portable Electronics (tr)
> **Case Type:** Electrical Engineering

Fact Pattern

Passengers board a standard-size plane carrying hand luggage and electrical devices such as Walkmen, beepers, CD players, and laptop computers. Business travelers quickly sit down and tap away on their laptops. A teenager turns on his Gameboy with headphones, while a college student listens to his favorite rap lyrics. Almost everyone on the plane is oblivious to the safety briefing given by the flight attendants.

The attendants walk down the aisles to check if everyone has safety belts fastened, seats upright, and all portable electronic devices (PEDs) off. A passenger sitting by a window with a laptop is told to stop using the computer before the plane takes off, but he says to himself, "I am almost finished with this ethics presentation. I'll finish my last three lines after she leaves. One laptop computer can't do any harm." Meanwhile, the flight attendant thinks that the computer is off when he closes the case. She asks the college student, "Sir, can you take your headphones off?" Seeing the guy still using his laptop, he says to himself, "What's wrong with this woman? Why she didn't tell that man in the front to do the same?" But he turns off his machine and removes the headphones.

In the cockpit the captain notices that certain displays on the instrument panel are slightly off the norm. He reports the error, then decides to proceed with the scheduled takeoff time. His departure is approved.

The plane takes off while some PEDs are still on. As the captain increases the plane's altitude, one of the compass displays is *far* from the norm. Knowing the flight routine, the captain uses his own judgment to get onto the correct course. The captain informs the flight attendants to make sure all portable electronic devices are completely off. The flight attendants notice that five laptops, along with a Walkman are still in use. The captain's readings are normal after the passengers actually turn off the PEDs and don't turn them back on until the plane reaches a certain altitude.

Questions for Discussion:

1. What are the facts in this case?

2. Are there any ethical questions?

3. Should the safety briefings be reconstructed? Why?

4. Does the captain have any idea why the readings are off? What should he have done?

5. Should air traffic control (ATC) ban PEDs in airplanes? If not, when should they be used during the flight?

6.8   Case Title: The Easy Way Out (rm)
Case Type: Electrical Engineering

Fact Pattern

Tony is an electrical engineer working full-time for LAN Corporation, which designs local area networks (LANs). Tony works approximately sixty hours per week, including weekends, and is pursuing his graduate degree part-time at the University of the United States; he is currently enrolled in two classes. His company requires him to maintain at least a B average in order to receive reimbursement for his education.

One of Tony's teachers has assigned an LAN-design project that will comprise 50 percent of the course grade. Recently Tony has been so busy with his job that his performance in the two classes has suffered and he has fallen behind. Tony's friend at work is an LAN expert who most likely could complete his project for him in under five minutes. This would make life much easier for Tony, since it would certainly help him receive a B average or better in the course and also provide more time for him to focus on his job and social life.

Questions for Discussion:

1.   What are the facts in this case?

2.   What are the ethical issues?

3.   What are Tony's options?

6.9   Case Title: But He's My Best Friend (nn)
   Case Type: Electrical Engineering

Fact Pattern

Gary is a microprocessor designer for TNT, a leading manufacturer of integrated circuits and semiconductor electronics. Last week, he was promoted to group leader and given the task of developing a processor that is to be the next-generation x-86 competition to Intel. Just before embarking on the project, Gary takes a week of vacation and visits Dave, his best friend since college. Dave is an electrical engineer with Kansas Instruments, TNT's main competitor. Gary arrives in Dave's hometown and decides to tell him the good news about his new promotion and project. However, just before he does, Dave describes to him the project *he* has been working on. Sure enough, Dave is nearing completion on Kansas Instrument's version of the P-7 chip.

Now, Gary feels he has a problem. Dave wants to show him the fruits of his labor, but Gary thinks that looking at the blueprints would be a type of corporate espionage. However, Dave and Gary have always talked about their work. Moreover, looking at the design would make Gary's life much easier, especially since this is the first big project for which he has full responsibility.

Should he tell Dave that he is working on a parallel project? Dave and Gary copied assignments from each other throughout their collegiate careers, but now things are a little different. He is afraid Dave will get upset if he takes the ideas he has worked so hard on. Also, since they are designing rival products, one is bound to be more successful than the other. The success or failure of their efforts has sometimes strained their friendship, which has lasted over ten years. However, he doesn't want to lie to his best friend about the job he is doing.

Questions for Discussion:

1. What are the facts in this case?
2. What are the ethical issues in this case?
3. Should Gary tell Dave about his job and/or project?
4. Is it ethical for Gary to look at Dave's design?
5. What would you do in Gary's situation?

6.10   Case Title: Honorable Minister
Desires . . . (ahs)
Case Type: Electrical Engineering

Fact Pattern

Bangladesh Power Development Board (BPDB) is a semi-government organization under the ministry of electricity, government of Bangladesh. It is responsible for development of power generation as well as its transmission and distribution. Pacific Consultant Ltd. (PCL), with long experience in power system engineering, has been working as a prequalified consultant to BPDB. Development International (DI) is a Canadian engineering firm with fifty years of experience throughout the world. PCL and DI have jointly entered into a contract with BPDB for all the engineering required to build a new steam power plant in Bangladesh.

As a part of planning and design, PCL and DI have surveyed the existing electrical loads and those predicted for twenty years; they have collected all necessary data for load flow studies. After those studies, and after considering the load center, communication and transportation facilities, supply of water, fuel source (gas), and other environmental factors, they select a site for establishment of a new thousand-megawatt power station. The proposal for the new site is submitted to the ministry through BPDB.

One day, Mr. Kader, project coordinator of PCL, who is responsible to the client, receives a telephone call from the chairman of BPDB informing him, "Honorable Minister desires that the site shall be shifted by three miles, to be near his native village, Taojan."

Mr. Kader, accompanied by Mr. Hepburn, the Canadian team leader, visits the new site and finds some disadvantages. It's hilly, far from both river (for water source and transportation) and highway. Transportation of heavy equipment would be difficult. Only the fuel source (gas) is available. Mr. Hepburn assesses that the new site will elevate the total estimated cost of the power plant by 10 percent. The operation and maintenance costs will also be affected.

Mr. Hepburn, team leader of an internationally reputed firm, does not want to recommend the new site. But Mr. Kader, project coordinator, is worried that, if they don't, both companies will lose their contracts or face difficulty in receiving future contracts with BPDB.

Questions for Discussion:

1. What are the facts in this case?

2. What are the ethical issues in this case?

3. What are the possible solutions?

4. What would you recommend in such a case?

> ### 6.11   Case Title: Broadway or Bust (aw)
> ### Case Type: Electrical Engineering

Fact Pattern

Knob Electronics is a small company employing approximately thirty people. Their primary function is the design of motor control distribution systems, which are used for raising and lowering backdrops for shows and concerts. Recently, the company has been having financial difficulties and is running the risk of going under if business doesn't pick up soon.

Michael Hunt, the owner of the company, decides that the only chance Knob Electronics has is winning the contract for one of the new Broadway productions. Being able to bid below all the other companies for the job, Michael lands the contract. Production is running smoothly, and they are having no problem completing the project in the time allocated. They are in the final phase of the design when things took a turn for the worse. Benjamin Dover Bright, in charge of shipping and receiving, notices that not all the cam locks ordered have been shipped. Knowing that all of them are going to be needed to complete the design, he calls the supplier, who says that all the cam locks have been shipped out; some must have been lost. He could ship out more, but this will take two weeks. Benjamin informs Michael of the situation, and they sit down to discuss their options.

They envision a few possible scenarios. First, they could substitute *used* cam locks, but there is a good chance they'll break down. If they break during one of the performances, the backdrop could fall, causing serious injuries. If the show management finds that inferior equipment has been used in the design of the motor control system, Michael could be sued. On the other hand, if the cam locks work fine and/or if nobody notices, his business will be saved. Finally, Michael could try getting an extension to wait for the new cams to come in, but that would certainly mean a decrease in the amount they would be paid. If that check isn't enough money, Michael might lose his business.

Questions for Discussion:

1. What are the facts in this case?
2. What are the ethical issues?
3. What would you do in Michael Hunt's place?
4. What other options does he have?

7

Mechanical Engineering

7.1 Case Title: Some Smoke (ra/san)
Case Type: Mechanical Engineering

Fact Pattern

Curts Inc. is a company specializing in power plants. Recently, their business has taken a jump forward, and the company has started obtaining contracts faster than they can hire and train individuals.

John is a mechanical engineer for Curts Inc. Because of the recent lack of employees, he has been juggling two or more projects at the same time. One project that he neglected for a while is a contract with the local utility company to modify an existing burner to use a more economical fuel and still produce the same amount of British thermal units (BTUs).

Now, the project is near the end, and he just got started on his assignment of giving a detailed report about the fuel to be used. Initial information was gathered by a team of engineers who concluded the fuel will be easy to obtain if imported from South America and that it will produce the same amount of BTUs. John finds out that the fuel, which is a very crude (tarlike) oil, is suspected of being carcinogenic when burned, according to a study done on this type of fuel. When he brings this to the attention of his peers, they accuse him of trying to destroy all of their work at the last minute.

John takes a moment to think over his options: Should he destroy the whole team's work, or forget about that *one* study and finish the project?

Questions for Discussion:

1.   What are the facts in this case?

2.   What is the ethical problem John faces?

3.   Does John have more than two options?

4.   Which course of action would you recommend to John?

7.2   **Case Title:** Whose Fault Is It? (kf)
       Case Type: Mechanical Engineering

Fact Pattern

Fabrication Inc. designs blades for steam-run power plants; these blades go as high as twenty-four feet in one direction. A nuclear power plant bought hundreds of these blades from Fabrication Inc. for a price well into the hundreds of thousands of dollars. Fabrication Inc. guarantees their product. After one year, several of the blades that were fastened and welded, cracked. The fabrication company is called in to inspect the cracks as part of their guarantee.

After the engineer for Fabrication Inc. inspects the parts, a meeting is called at the fabrication company. The main speaker is the inspecting engineer, Bob, who explains his findings to three other engineers and the manager. Bob states that the cracked blades are the fault of Fabrication Inc. All of the blades, with and without a visible crack, have been welded incorrectly. Due to this improper welding, when stress is applied to the blade, it starts to crack at this weak point.

One engineer, John, suggests that, instead of living up to the guarantee and losing an enormous amount of money, Fabrication Inc. should state that the cracks were caused by the poor steam quality of the power plant. Fabrication Inc. could then offer, for a substantial price, to repair their blades and place an extra coating on the blades as a protection against the steam. This way, Fabrication Inc. can repair all the blades and not lose any money.

Another engineer, Paul, states that the power plant's engineer should investigate the problem, but Fabrication

Inc. still should maintain the position that John suggests. But the plant should get a substantial *discount*, since the problem is ultimately the fault of Fabrication Inc.

A third engineer speaks. Frankie believes that Fabrication Inc. should own up to its mistake because those blades should have been checked before they were sent out. What if they do not give back to Fabrication Inc. *all* the blades but just the ones that visibly are cracked? Suppose the power plant decides not to have the others coated because they will have their steam quality checked and upgraded? Someone could get hurt or killed. The safest position is to admit their mistake, take the loss, and create loyalty in this partnership.

Taking all the information, data, and suggestions into consideration, the manager, Kevin, decides to follow Frankie's advice. The company ends up losing over two hundred thousand dollars. The CEO of Fabrication Inc. is dissatisfied with Kevin's decision and fires him.

Questions for Discussion:

1.   What are the facts in this case?

2.   What ethical dilemma has occurred?

3.   In what way is the company to blame for the cracks in the blades?

4.   Is the power plant responsible in any way?

5.   Is the company CEO justified in firing Kevin?

6.   Is there any other action Kevin could have taken that would have allowed him to do the right thing without being fired?

> **7.3   Case Title:** Not Exactly Child's Play (kf)
> **Case Type:** Mechanical Engineering

Fact Pattern

A company called Baluga is known for designing toys. Bill, an engineer, has been asked by his Baluga boss, Don, to build a surveillance camera small enough to fit in an overhead sprinkler, with a wide view lens that would encompass a table. The boss also asked Bill not to tell anyone what he is working on, for security reasons. When Bill asks Don for what purpose the camera will be used, the boss indicates that such information is released on a "need to know" basis, and Bill does not need to know.

He is given no limit on his budget for this new product, which is unusual for Baluga. Bill designs the camera and has it ready in a couple of months.

One night, when he has stayed at the office late trying to work out all the bugs in his prototype, Bill overhears Don and his colleague Joe from his cubicle. They are talking about how they have embezzled money. When Don tells Joe that someone in the company is onto them, Joe starts to panic. Don calms Joe down by telling him that he has Bill working on designing a camera small enough to fit in an overhead sprinkler; they can frame someone in the office to take the fall for the embezzlement by tricking an employee into transferring accounts around and catching him on tape. Joe suggests that they open up an account for this person and place some money in it for him.

As Bill hears all of this, he quietly picks up all of his belongings and leaves for the evening. Bill does not know if he should stay quiet because they might make him the fall guy. Or, if he stays quiet *and* blackmails his boss, Bill could keep his job *and* maybe get a raise. In the end Bill decides not to finish building the camera. For the next several months he tells his boss that there are some electrical problems with the camera that need to be worked out. Bill's decision not to finish the product saves his job and allows his boss and colleague to be caught.

Questions for Discussion:

1.   What are the facts in this case?

2.   What ethical dilemma occurs?

3.   Is Bill responsible in any way, even if he's done nothing wrong?

4.   *Should* Bill have told someone?

5.   Would Bill be responsible if his boss and colleague did not get caught?

6.   Is there any other path Bill could have taken?

7.4 Case Title: What Were You Thinking? (kf)
Case Type: Mechanical Engineering

Fact Pattern

A company called Forced Air develops hair dryers. John heads the design of the latest hair dryer, which delivers nine hundred watts of power, the most any hair dryer has ever produced.

Paul, a factory worker, is given several different instructions for testing the product, a process that takes several months. The hair dryer passes all tests but one, which studies how the dryer reacts when it burns out due to excessive use; one out of five hundred blow dryers catches on fire and melts during the burnout phase.

Paul reports his finding to John. John reviews the results and concludes that the product is ready to be marketed; he decides against redesigning the product to work out the bugs, which would take too much time and money. The design is already over budget and taking place on borrowed time; John has given his word that he will have a working product by the end of this month.

John gives the public relations and marketing departments the OK to market this product; they spend months as well as hundreds of thousands of dollars on advertisements, publicity, et cetera.

Six months after the product is on store shelves, an eleven-year-old girl loses the sight in one eye, and her face is disfigured badly, with third-degree burns, after using one of the dryers. This incident is reported to the company, but John explains it as a freak accident to his bosses.

After repeated accidents and a class action suit because of the flawed design of the new blow dryer, the company recalls all of them. The company loses a great deal of money because of John's "go-ahead" decision, and a lot of people get hurt because of his negligence.

Questions for Discussion:

1. What are the facts in this case?

2. What ethical dilemma occurs?

3. What are John's responsibilities, if any?

4. Is there any other path that John could have taken?

5. Is the company responsible in any way?

6. Is there any other path that the company could have taken?

> **7.5 Case Title:** Light for Free (ag)
>   **Case Type:** Mechanical Engineering

Fact Pattern

Joe has just bought a summer house on the island of Sicily, a villa half a mile from the calm, blue waters of the Mediterranean Sea. Joe is a talented mechanical engineer and very handy around the house; he plans to do much of the mechanical and electrical work on the villa by himself.

While he is installing a Jacuzzi in the back of the house, Joe discovers that he can connect the electricity from the villa to the power lines coming from the town's electric generator. He devises a method to make the modification without affecting the power to the rest of the town.

The advantage of doing so is obvious: Joe will not have to pay any electricity bills while he lives in his summer home, and, as far as he knows, the hookup cannot be traced easily back to him. However, Joe has no data on whether the town generator can handle the extra electric load. He figures he is going to be living there only one month out of the year; how much damage can he cause?

Questions for Discussion:

1. What are the facts in this case?

2. What is the ethical problem with what Joe is doing?

3. Do you think that Joe is smart in what he's doing?

4. Would you, given the same circumstances, do what Joe does?

5. Do you think it is a smart choice to make the hookup?

> ### 7.6   Case Title: Final Exam (ag)
> ###   Case Type: Mechanical Engineering

Fact Pattern

Tom is preparing for his final exam in thermodynamics, a course that he has been struggling with all semester long. He desperately needs to pass this exam because as it stands he has a D in the course; not passing this final exam means he will have to repeat the class.

One night Tom sees a janitor that he is acquainted with, and they begin to talk about his course difficulties. The janitor, whose name is Mike, likes Tom and thinks he's a good kid. Mike offers to help Tom with his thermodynamics exam.

The student, bewildered as to how the janitor can help, asks, "Did you take thermo while you were in school, Mike?"

"No, Tom, I went only as far as high school, but I know for a fact that your professor keeps all of his exams on his desk; I saw them when I was cleaning his office one night."

Mike says he can let Tom into his professor's office to look around for the final exam. Tom is excited about the idea of getting it ahead of time and feels a sense of relief.

On the other hand, Tom realizes that if he gets caught, he can get thrown out of school, and that would not be good. He has two options: to study hard for the exam (and possibly still not do well) or obtain the exam (at the risk of getting caught). Which is the best alternative?

Questions for Discussion:

1. What are the facts in this case?

2. What is the ethical problem with what Tom is doing?

3. Do you think Tom should try to get the exam ahead of time?

7.7 Case Title: Safe in the Sun (ssh)
Case Type: Mechanical Engineering

Fact Pattern

Ricardo is beginning his fourth year as an engineering student at a university in Boston and is now in charge of a team of students who are determined to build the school's first solar-electric car. In order to gain corporate donations and university funding, the car must gain a great deal of exposure. It is decided by the team that the car will race in the Tour de Sol in May 1996. This gives the students a goal and provides ample justification for investment.

The time frame from the team's inception of the idea to the date of the race is less than a year; this fact poses the obvious problem of feasibility. To overcome this difficulty, the group presents the solar car project to all of the senior design classes in every discipline of engineering. The idea is to split the car into many smaller projects; teams would finish their concentrated assignments and present them as senior projects. The idea is also pitched to marketing, business, and communications seniors, who can design investment solicitation plans and help the team gain access to local and national media.

Within the next few weeks, teams are set up for all known aspects of the project, and Ricardo is to coordinate everything—creating communication links between teams, setting goals, and negotiating timelines for the integration of all the projects. Key to all aspects of the project are safety considerations. In addition to all government safety regulations, the car has to abide by several safety guidelines set by the Tour de Sol. In order to obtain money and materials, the team has to convince every sponsor that all Tour de Sol requirements will be met on time.

Over the following few weeks, car body designs are submitted, altered, and reworked until a final design is agreed upon. At that point a massive effort is made to obtain the materials and components necessary to construct it. With those in hand and a procedure plan laid out, groups of five to twenty people work in rotation twenty-four hours a day for the next month to complete the monocoque of the car (the shell made of all advanced composite materials). At this point every team has the huge task of integrating the final design of its product to meet the actual specifications of the monocoque and the preliminary specs of other product groups. During this process the interior design group discovers that miscommunication between their group and the body team has led to a monocoque design that is not tall enough at certain critical points. There is a specific regulation regarding the height from the bottom of the lowest part of the seat to the ceiling of the car. The height of the car at this point is about an inch too short, and the seat configuration also takes an additional, unexpected inch of height. This causes a two-inch disparity between the actual and the designed configuration, and a breach of regulation by about an inch. More important, this change also directly reduces the distance between the top of the driver's helmet and the ceiling, which means that, in the event of even a minor crash, the driver's chances of serious injury to the neck are increased. The interior group analyzes the situation and determines that the seat angle cannot be changed to reduce the height because they have designed the seat based on the largest ergonomically feasible angle to minimize the height in the first place.

A meeting is held including Ricardo and all of the team members involved in the process. It is determined that there is nothing that can be done in the interior of the car without breaking several safety regulations directly related to the driver. The only possible way to bring the car within these regulations is to increase the height of the monocoque, which will require the reconstruction of the entire monocoque and is impossible from both time and resources standpoints. Ricardo is faced directly with either enforcing a redesign that would effectively scrap the project or keeping a design that breaks the height regulation. He decides to postpone the decision by allowing everyone to work as scheduled and assigning a person from each of the two relevant teams to try to find an alternative solution. Weeks pass, and the deadline is approaching. The two team members report that there is no way, outside of a new monocoque, to change the height without causing more severe problems. They do report, however, that this regulation was never actually checked by a third party in past races and is used more as a guideline than as a hard-and-fast rule. Since there seems no alternative and apparently no punishment is rendered for breaking the rule, all of the group members agree to allow the "violation" to go unchanged. Ricardo despises the predicament and the decision, but does not want to let his entire project go down the tubes for a technicality. The decision holds and the car is finished barely in time, yet Ricardo is still uneasy in the final days before the race.

Questions for Discussion:

1. What are the facts in this case?

2. What is the ethical problem facing Ricardo?

3. How could Ricardo have anticipated the problem before it was too late?

4. What are Ricardo's options in handling the situation?

5. What do you think should be done?

6. What do you think Ricardo's actions should have been upon learning of the situation?

7.8   **Case Title:** Build an Arm to Be a
Leg Up on the
Competition (ssh)
Case Type: Mechanical Engineering

Fact Pattern

Inaki is in his last year of college, studying mechanical engineering. One of the last classes he has to take to graduate is a senior design project class. He and a few other students have to pick a project idea and develop it to create a final product by the end of the semester. The group decides upon an arm that will attach to a desk and hold a computer monitor, allowing the monitor to be raised, lowered, or rotated anywhere within 360 degrees. The arm could also be extended, which would allow a user to sit almost anywhere in the room, even on a bed, and work at the computer.

This idea excites Inaki and the rest of the group a great deal. Everyone contributes in creating a very practical yet versatile design. When the group presents their design in class, it is very well received. The response from other members of the class and the professor leads the group to consider actually marketing and manufacturing the product when the project is completed.

As the semester draws to an end, the group finally finishes every aspect of the design to—in their opinion—perfection. A prototype is built of plastic, and although it is an excellent depiction of the design, it is used only to show the *concept*; a great deal more work is necessary to build a prototype that will actually support the weight of the monitor at any position the user may choose.

Inaki is very proud of the work that he and the group have accomplished and is not satisfied completely with the A that he receives from the class; he is determined to actually produce the computer arm. All of the students graduate, and only Inaki is interested in taking the project further. During the summer he invests money to purchase materials and lays out a plan to have a contractor manufacture the product. A comprehensive stress analysis has been done for each component of the product; however, the actual manufacturing process has not been taken into consideration because it is not fully understood. Inaki hands the design drawings over to a contractor, who builds each of the sections from steel as is called for in the design. The base of the arm is shaped like an L; it was decided earlier by the group to make the base from one piece instead of two to make it stronger and possibly cheaper. All of the calculations are based on the original strength of the material, assuming that the properties of the steel will not change when formed into the arm itself.

The contractor, with Inaki's approval, makes the L shape by heating the steel and bending it to fit the design. The contractor is not aware of the specific strength requirements and is only working from the drawings given to him. It turns out that the bending process causes the steel to harden and become more brittle. Also, occasional micro cracks form, a process that can reduce significantly the ultimate strength of the material at the bend, where pressure is greatest.

Over the next couple of months the manufacturing process is completed, from general material forming to assembly all the way to packaging; however, the possible problems with the bend go unnoted and uncorrected.

Questions for Discussion:

1.   What are the facts in this case?

2. Who is responsible for the product's flaw in the case of an accident?

3. Could/should Inaki have anticipated the problem?

4. Were the contractor's actions acceptable and responsible?

5. Is this "mistake" acceptable?

6. Is it too late for the problem to be corrected if it is detected?

7.9   Case Title: There's No Such Thing
            as a Free Seminar (rp)
      Case Type: Mechanical Engineering

Fact Pattern

Bryan is the director of the radiology department of a major metro-area hospital. Through some trade publications, Bryan learns of a new type of imaging system being developed that combines a computerized axial tomography (CAT) scan with a virtual-reality setup; this combination would allow a physician to "fly" through a patient's colon, for example, and look for any abnormalities.

Several competing manufacturers, including New England Imaging (NEI), are developing these systems. The aging imaging equipment at Bryan's hospital is due to be replaced in the near future, and Bryan contacts NEI to request additional information on their new system.

Within several days an NEI representative contacts Bryan and tells him that due to his professional status, he can attend an upcoming two-day seminar on the "virtual CAT scan" in a posh midtown hotel and his registration fee will be waived.

Bryan is no fool; he realizes that this "educational" seminar will most likely be a big sales pitch for NEI's system. However, it would be nice to get away from the often-hectic pace of the hospital for a couple of days. Besides, all new medical equipment that is to be purchased by the hospital first must be evaluated by the clinical engineering department; this process involves researching systems from other manufacturers as well as getting feedback from the hospital staff members who use the equipment. Bryan will have *some* input when the

final decision is made, but he feels that he is too professional to be "schmoozed" into advocating a system that
may turn out to be problematic.

Questions for Discussion:

1. What are the facts in this case?

2. What is the ethical issue?

3. Should Bryan attend the seminar?

> **7.10   Case Title:** It's Our Only Chance (rp)
> **Case Type:** Mechanical Engineering

Fact Pattern

Keith is the president of Advanced Pulmonary Devices (APD), a small Connecticut-based medical device company that has developed a new type of ventilator. This device uses a technology known as high-frequency oscillatory ventilation (HFOV) and has proven to be effective in cases of neonatal respiratory failure where conventional ventilation has been unsuccessful. On June 26, 1993, the FDA Advisory Panel for Respiratory Products recommended that the device be approved for treatment of neonates with respiratory failure, and APD began commercial distribution.

APD has since been working on a modified version of their ventilator that would be used on pediatric and adult cases. Throughout this process Keith has developed close alliances with many nearby hospitals and physicians; some are using APDs device in their neonatal ICUs. Development of the modified ventilator is progressing well, and preliminary studies have shown significant benefits in improving oxygenation and in survival of chronic lung disease. Keith has every reason to believe that FDA approval for the new ventilator is not far away.

Late one evening Keith receives a call from Dr. Weismann, a physician at a nearby hospital. A pediatric patient has been brought into the unit's ER in acute respiratory distress and is not responding well to ordinary ventilation. Dr. Weismann knows of the new ventilator and believes that this device could provide lifesaving treatment to this patient. He asks Keith to provide him with one of the units, with the agreement that he (Dr. Weismann) will assume all liability.

Questions for Discussion:

1. What are the facts in this case?

2. What is the ethical issue?

3. Should Keith provide Dr. Weismann with one of the ventilators?

4. What do you think would happen if the new treatment failed? What if it is successful?

7.11   Case Title: It Won't Happen Again (rp)
Case Type: Mechanical Engineering

Fact Pattern

Scott is the director of the clinical engineering department of a mid-sized, metro-area hospital. His responsibilities include managing a team of two engineers and three technicians, whose jobs are to maintain and repair all of the medical equipment in the hospital. Although he feels his department is understaffed, they still manage to keep up with the workload and have a very good reputation as a whole.

Within the span of three months, two of Scott's employees leave the hospital. Working with a skeleton crew, the department manages to keep up with repairs but begins to fall behind with the scheduled preventive maintenance. Months pass without either vacated position being filled, and the preventive maintenance forms begin to pile up. This situation is made more critical by the fact that the Joint Commission of American Hospitals (JCAH) is due to inspect the facility in several months, and preventive maintenance records are typically scrutinized during these inspections. If the inspection goes poorly, the JCAH might refuse to accredit the hospital.

Scott interviews several candidates and comes across one that he thinks might work out. An offer is made and the candidate, Jim, accepts. From the start, Jim's work habits seem questionable: He appears forgetful, and, although the quality of his work is generally good, his productivity level leaves something to be desired. Over time he starts to become more productive, however, and he begins to make some headway with the overdue preventive maintenance.

But when Scott takes a day off, he receives several complaints from the hospital staff upon his return. Jim seemed to be "out of it" the previous morning, and the question of a possible substance abuse problem is raised. Scott's initial impulse is to lay Jim off.

The hospital is poised to announce another large layoff, however, and Scott is afraid that if he lets Jim go, Jim's position will be closed. Even if the position remains open, he realizes that it most likely would not be filled before the JCAH inspection takes place. With such a staff shortage, it would be nearly impossible to catch up on preventive maintenance.

The other employees in the department have also noticed Jim's erratic behavior and confront him about it. Jim admits to them that he has a problem and will make every effort to keep it under control. Two weeks pass, and Jim's work remains satisfactory. Scott has yet to approach Jim about the "out of it" incident, but is aware that the rest of the department has already done so.

Questions for Discussion:

1. What are the facts in this case?

2. What is the ethical question?

3. What issues are at stake here?

4. What course of action do you think Scott should take with Jim?

7.12   Case Title: Ties That Bind (cr)
          Case Type: Mechanical Engineering

Fact Pattern

Antonio is a mechanical engineering major at a state university. Anticipating graduation in May, he has been interviewing with several companies for a job. About three weeks ago, he was offered a job from PRIT Technology, as a junior systems engineer. His base salary would be thirty thousand dollars, with a one-time signing bonus of five thousand dollars and two weeks' vacation.

Antonio has a part-time job with CTP Inc., a small engineering consulting firm located in Brooklyn, New York. He has mentioned his desire to continue working for their firm upon graduation; however, they have not made him an offer yet.

When PRIT Technology visits the campus to recruit other job candidates, Antonio makes a verbal agreement to sign with their firm. He shakes hands with the recruiter, who agrees to send him the formal papers at a later date; Antonio did *not* sign any papers tying him to work for their firm.

CTP Inc. recruiters have just started their signing process and are considering Antonio, who is a good worker. To train another person would take up valuable time and money. Therefore, they make Antonio an offer of forty-two thousand dollars, with a five thousand dollar signing bonus and paid vacation of one month.

Now Antonio has a dilemma. He really likes working for CTP Inc. and wants to continue there. But since he has made a verbal agreement with PRIT Technology, he feels somewhat obliged to work for them. However, his familiarity with CTP *and* the extra money they're offering make theirs a job that's hard to refuse. What should Antonio do?

Questions for Discussion:

1. What are the facts in this case?

2. What are the ethical dilemmas in the case?

3. Is Antonio's verbal agreement a binding contract?

4. Is it wrong for Antonio to turn down PRIT and work for CTP?

Appendixes

A

Ethics Training: An American Solution for "Doing the Right Thing"

HAROLD J. TABACK

ETHICAL DECISION MAKING

Ethical decision making, by definition, is difficult. In fact, if the decision is not difficult, it probably is not a matter of ethics. Ethical decisions involve an element of self-sacrifice. The sacrifice might involve losing a client or a job. It can be a matter of denying your family because your job is at stake if you resist taking an action that would threaten public health or be a violation of the law. Issues come up unexpectedly, and often a person with the best of intentions makes the wrong decision. It often takes training to even recognize a moral dilemma. For example:

---

*Hal Taback Company, 378 Paseo Sonrisa, Walnut, California, 91789

You are a successful young project manager in a consulting firm that has national accounts with several major manufacturing firms. Because of your previous excellent performance, you are assigned to manage environmental assessments at several plants of one of these large companies. The contract is being conducted under legal counsel, and the results will be handled to maintain privilege under the law. You find a toxic chemical release that you know will clearly threaten public health. You contact the lawyer who ordered the study, and report your findings. The lawyer directs you to stop work at that site, to move on to another plant, and not to submit your findings in writing. You point out the need to report this to the responsible agency. You are reminded of the contract that provides for the confidentiality of the findings, and are told that, if there is to be any reporting, the lawyer will handle it. You are warned that this information and this conversation are to be held confidential and *"will remain between the two of us. Do not discuss it with anyone!"* Your attempts to learn how the lawyer will proceed on this information are met with the comment *"It is not your concern. You do your job [i.e., to investigate], and I will do mine. If I need any additional information, I will ask for it."*

You have no knowledge as to when or if the release will be reported. From the conversation with the lawyer, you know that any attempt on your part to learn further details on the disposition of your information will be met with hostility. Should you follow orders and go to the next plant? Should you document the findings in a formal report? The client might not pay for this action, since you were ordered not to prepare the report. Whether or not the law in this state requires you to report the release, if some adverse health affect occurs, the injured parties could sue the manufacturing firm. Eventually, it will be discovered that you and your consulting firm were in possession of knowledge of the release. What if your client decides to not report it?

This is truly an ethical dilemma and requires some difficult decision making. If you never have contemplated such a dilemma, you might decide to follow the dictates of the lawyer. After all, lawyers are supposed to tell you what to do under the law. Besides, your firm signed a contract that says the data are confidential. You might feel, "OK, I will go to the next site and not make any waves."

If this scenario or one like it had been discussed in an ethics training session, you might have learned that your firm would advise that you elevate this situation to your supervisor, or further if necessary. Your management would want to ensure that some corrective action is taken to protect public health. Also in that training session, you would have learned that, as an environmental professional, your first obligation is to protect public health. You would know that to go on to the next plant is not the right thing to do. You would be prepared to do the right thing. The specifics of how this issue would be resolved in any consulting firm depends on the relationship of the principals involved. Some considerations will be discussed in the following paragraphs. But, if the consulting firm had held ethics training on a regular basis, management's position on situations like these would be known and could be implemented with confidence and diplomacy.

ETHICS TRAINING PROTOCOL

Ethics training, like any formal training, consists of theory and practice. Maybe it is more analogous to lecture and lab. It involves learning what is right and then practicing it on a regular basis. Overcoming the human tendency toward selfishness, which can have such a strong influence on ethical behavior, requires just as much practice and discipline as overcoming the body's physical limitations in achieving athletic success. One can read books and watch videos on hitting a baseball or golf ball, playing tennis, swimming, or performing gymnastics. But perfection comes only with repeated practice. Baseball players play nearly every day but still take batting practice before each game. Professional golfers often hit a bucket of balls before each round in a tournament. It is not sufficient to have a one-time lecture of dos and don'ts regarding professional behavior and expect the professional to perform correctly, to do the right thing in tough situations, when there is a risk to personal comfort and security as well as to that of loved ones. It is also insufficient for a firm to produce a code of ethics and have employees sign a certification indicating that they have read and understand that code.

The important part of ethics training is the periodic discussions of case studies, real or hypothetical. Since discussions of this type seldom occur spontaneously, they must be planned and conducted on a regular basis. The first few sessions would cover basic values, the obligation of the environmental profession, management's position, and the firm's *modus operandi* for dealing with ethical issues. From then on, the sessions should concentrate on case studies evaluating alternative solutions, using the value system and *modus operandi* learned previously.

The recommended format for the ethics program is to work within suitable company groupings. The preferred arrangement is such that the people in the group are in the same area so that periodic one-hour sessions could be convened, the same as any routine meeting for communicating with employees. Ideal timing for these discussions is during regular business hours. However, sessions conducted over the lunch break, often referred to as "brown-bag seminars" are effective if employees willingly agree to participate in this manner. (It works best if the employer provides lunch.) The corporate ethics trainer would train a facilitator in each group, provide training material, and observe the individual workshops to help improve quality. Facilitator training would consist of bringing together those who volunteer or are selected to facilitate with the trainer, who would conduct a workshop similar to those that the facilitators would eventually conduct themselves. In addition to the information discussed in the previous paragraph, facilitators would learn how to facilitate without preaching to attendees. They would be shown how to deal with strong individuals who might try to dominate sessions and how to draw out timid participants sensitively.

Ideally, the group manager and top management should appear at the initial sessions of the facilitator group and the individual groups. They should state the company's position and inform the employees of their duties and responsibilities in regard to doing the right thing. When possible, a day-long seminar with top management should be held to acquaint them with the program, review the values being discussed, and reach a consensus on the modus operandi to be used when an ethical dilemma arises. The trainer can act as the facilitator in this top management meeting after presenting his/her introduction to the program. However, some thought might be given to using a third-party facilitator for the meeting, since the trainer will be somewhat of an advocate in this setting.

If the firm is large, with many individual groups, the introductory presentation by top management can be accomplished by video. The trainer would work with

corporate staff to prepare a presentation outline and ultimately the scripts. This video will be a valuable statement of the company's concern that all of its employees
recognize the issues and do the right thing. An effective
environmental compliance program, one including values and ethics training, is recognized by the *United States
Federal Sentencing Guidelines.** In the event of an inadvertent environmental law infraction by an employee, this
training video could provide convincing evidence of
management's commitment to the environment and
could lead the U.S. Department of Justice or the Environmental Protection Agency to forgo criminal prosecution
according to the Guidelines.

CASE STUDY

Having established the responsibility of the environmental professional to protect public health, we can explore
how it applies in specific cases. Consider the case
already described. The young project manager has just
ended the initial phone conversation with the client contact, the lawyer. If s/he has had ethics training, s/he
might use the values checklist that follows before elevat-
ing the situation to management. It is most important to
be sure of the facts. Could there be any error in the data?
Have verification tests been run? Should an uninvolved
colleague be consulted to review the findings before anyone goes to management? Clearly, the confidentiality of
the data must be maintained. However, you cannot
afford to be in error when so much is at stake.

---

*Wilcox, Reynolds, and Theodore, "U.S. Federal Sentencing Guidelines and the
Development of Ethics Education Programs in the Environmental Industry,"
paper 96-TP161A.05, presented at the 89th Annual Air & Waste Management
Meeting, Nashville, Tenn., June 23–28, 1996.

After all has been checked and the necessary analysis conducted to indicate the certainty of risk to public health, *before* the call to the client, it is time to consult management. If this checking is still under way and will take some time, then management must be advised and informed of the verification plans. It is unacceptable for management to learn of the situation for the first time in a phone call from the client. Ideally, a management team meeting will be held at which the project manager will lay out the facts. The communication link between the consulting firm and the client will be decided, usually at the highest level in each firm that has an established relationship. The necessary action should be taken to report the toxic pollutant release. The project manager should participate to the extent the client wishes but should ensure in any event that s/he is kept aware of the proceedings.

At times like these, it is important to maintain respect and mutual concern. The client's attorney may seem dictatorial initially and the project manager may feel some hostility. However, after the two managements have resolved a plan of action, the project manager and attorney probably will need to work together. Therefore, it is important that the project manager maintain a respectful demeanor and an attitude of competence in dealing with the attorney from the outset. Human nature can cause some of us to react spontaneously to difficult situations, thereby upsetting a business relationship. The situation described in the case study could well be one of those. However, if the project manager has already been exposed to a hypothetical situation such as this in an ethics workshop, s/he might be reminded of the need to practice restraint and to maintain dignity.

In most situations a result satisfactory to all parties can be achieved. However, in the event that your management does not respond appropriately, *you are not relieved of your responsibility to do the right thing.* You must pursue the proper action in a dignified manner, keeping in mind the values of justice and fairness. Try to understand the position of those with whom you disagree; try to convince them with facts and reason. Protecting the public from exposure to toxic pollutants is primary, but only if a more adverse impact is not created. Such a possibility could involve immediate physical and emotional stress caused by the loss of jobs when a plant is closed after severe penalties are imposed as the result of an inadvertent release by a confused employee. Compassion for families that might suffer due to plant closure is not inconsistent with doing the right thing. Each individual must establish her or his own values interpretation, which in some cases will dictate leaving a firm because of a disagreement in values. Before one makes a decision of this magnitude, it is important to use a checklist such as the one shown to evaluate the issues involved.

VALUES CHECKLIST

This values checklist requires some serious contemplation and reflects the thoughts of the author. It is neither complete nor inherently correct. It is suggested that the reader use it as a point of departure to examine all sides of an issue and that this format be used whenever an individual is faced with a professional ethical dilemma. After consulting or completing such a table, you should take your contemplations to a colleague, supervisor, or, better yet, a group of confidants for a thorough discussion.

VALUES CHECKLIST FOR THE CASE STUDY

Value	Issue	Action
Trustworthiness		
Honesty	Accurate & thorough reporting & interpreting the analytical data & answering all questions. Giving appropriate credit to participants in the report.	Disclose the findings; do not conceal questionable details; hold accurate but unrestrained dialogue. Involve your staff.
Integrity	The courage to report the adverse findings, realizing the possible consequences, and to advise client in advance of this intent.	Report promptly to the client and to your supervisor. (The case study does not indicate that firm notified client of intent to report.)
Sincerity	Showing concern and appreciation for the seriousness and possible consequences.	Maintain a serious and concerned attitude in all dealings with the client.
Loyalty	Looking out for your firm's interest in this delicate position.	Inform your supervisor & accept his/her participation in the disposition.
Promise Keeping	Meeting the contractual requirements both explicit and implicit.	Maintain the confidentiality of the assessment data.

(continued)

275

Value	Issue	Action
Respect		
Courteousness	Politeness and respect in all interaction with clients, supervisors, colleagues, & subordinates.	Interact pleasantly with client's lawyer even when in disagreement.
Punctuality	Being on time for meetings, meeting program due dates, etc.	Not a particular concern in this case, but applies across the board.
Right of Self-Determination	Respect for each individual's right to decide for him/herself, even if wrong.	In dealing with the lawyer, understand and respect this right.
Responsibility		
Pursuit of Excellence	Maintaining knowledge of the latest in technology and proficiency in the tools of your profession.	Insure that the release detection reflects best measurement technique.
Competence	Maintaining control of the situation from technical, safety, & management standpoints.	Demonstrate competence.

Integrity	The courage to *do* the right thing when the consequences may be adverse.	Insure that client (or responsible party) will execute cleanup as appropriate.
Self-Restraint	Considering facts and circumstances, consulting with supervisor & colleagues, before acting w/care.	Elevate the situation to management and facilitate a plan of action.
Justice and Fairness		
Open-Mindedness	Recognizing that there are different solutions to every dilemma. Willingly considering them. Offering to cooperate with independent assessment when the seriousness of the matter warrants it.	Try to learn and understand any workable plans of the attorney. Suggest client get a second opinion to verify findings.
Ability to Admit Errors	Recognizing when you are wrong and openly admitting it. Releasing work product for checking by an independent evaluator.	If there are any anomalies in the data, reveal them immediately. Get uninvolved colleague to check work.

(continued)

Value	Issue	Action
Caring		
Kindness	Helping others to achieve their legitimate goals.	Offer to help the attorney deal with situation. Explain risks & benefits.
Generosity	Offering money and free help and advice to a colleague for problem solving, etc.	Offer to meet with client on an off-the-record basis to discuss issues.
Compassion	Recognizing downside to employees, families, & other stakeholders of various actions & attempting to mitigate hardship.	In seeking resolution to this dilemma, keep in mind the impact on these stakeholders.
Desire to Avoid Harm to Others	Protect public health consistent with a concern for public welfare.	Continue to seek prompt resolution to mitigate health & welfare impacts.

Civic Virtue and Citizenship

Social Action	Public critiquing of regulations that overprotect.	Not directly relevant
Public Service	Holding public office & supporting political issues.	Not directly relevant
Opposition to Injustice	Using expertise to quell radical control, etc.	Not directly relevant

CONCLUSIONS

Ethics training is valuable to everyone. It sensitizes us to ethical issues and prepares us to respond appropriately to ethically questionable situations that arise, usually unexpectedly. Workshops that include management participation are especially effective because values can be shared and employees can feel comfortable about their probable reactions *before* dubious situations occur. Imagine a CEO who stands up in the workshop and says, "I have always respected the truth. You will never have trouble with me if you tell the client the truth and come directly to me with any conflicts." The employees in that workshop would have enhanced confidence in their employer's support for doing the right thing. In addition to reducing the risk of employees doing the wrong thing on the spur of the moment, these ethics workshops build confidence between management and employees that strengthens the relationship on all fronts.

The recommended format is to hold semimonthly or monthly seminars during which a moral dilemma is discussed. These discussions should follow a half- to full-day workshop at which the values suggested here are reviewed and discussed along with a presentation of management's recommended protocol for handling ethical issues.

B

Ethics Bowl: Applying Ethical Reasoning to the Professional World

ANDREW REHFELD

But on every subject on which difference of opinion is possible, the truth depends on a balance to be struck between two sets of conflicting reasons.

—J. S. Mill, *On Liberty,* Chapter 2

When Joe Camel was introduced ten years ago, who could have predicted he would be the intellectual fodder for four college students? Huddled together, these students are now responding to a simple question: Even if a tobacco company is legally allowed to use cartoons to advertise products, is this use morally justifiable when it targets minors as the consumers of tobacco products? The four juniors and seniors, seated behind a long table, whisper quietly but hurriedly, organizing their answer.

A clock ticks; the audience watches. A minute passes. The moderator asks for a response.

A spokesperson from the college team takes the microphone and announces her group's decision: As long as the law fully permits cartoon advertising, the cigarette manufacturer has no moral obligation to refrain from the use of Joe Camel. But, the student adds, the law itself is an unethical law. The manner in which a product is advertised should be commensurate with the kind of product available. Thus, while the company is morally allowed to advertise, the law itself should be changed to prohibit such advertising.

As the students finish, a panel of four judges begins questioning them. "Are you saying," asks one particularly incensed juror, "that every law should be followed without ethical impunity? Are you saying that it is only the legislature that decides what is or is not moral for society? What about laws that are manifestly unjust? Don't we have moral obligations *not* to follow them?"

The spokesperson responds, "Of course one should not obey an unjust law. For example, we don't think anyone should discriminate against a person because of race or gender, no matter what rights to discriminate may exist in law. But there is a difference between laws of commerce and those that force a person to do harm to another in the name of some immoral belief. There are different standards in the marketplace. The role of government is to create a level playing field for all businesses. If something is immoral, the government must not allow all businesses to act that way. But once government permits it, then it is not only *not* required for a firm to act thus but foolish for a firm to act in any other way."

"On what basis," another judge asks, "do you distinguish between commercial laws and, say, civil laws in terms of ethics? Isn't harm of any kind harm whether it comes from advertising or a blow to the head? If a law allows shop owners to kick any customer they don't like, just because they don't like him, should that law be followed? Why does commerce grant some exemptions from morality?"

The questioning continues and, as it does, the team argues strongly for the distinction between commercial and noncommercial life. After ten minutes of discussion, the moderator reads a prepared answer: The company acted unethically if it knew, or should have known, that advertising cigarettes through the use of a cartoon was primarily attracting minors to smoke. This "moderator's answer" is offered as a suggestion—one way of framing the debate. The team now has to respond. Do they agree with the moderator's answer, disagree, or do they agree with a modified version of the answer?

The members of the team huddle together, and after another minute the spokesperson responds. The group disagrees with the moderator's answer. But, they say, if the company knows that the advertising is encouraging minors, primarily, to smoke, it must lobby for legislation that bans such advertisement.

The judges take a moment and fill out their score cards. They rate the team on four separate scales of equal weight. First, consistency: Is the team's response intelligible, that is, is it stated and defended in a logically consistent manner? Second, ethical relevance: Does the team omit any ethically relevant points in their analysis? Third, ethical *irrelevance*: Is the team's response based on any irrelevant considerations? Fourth, reasonableness: Is the substance of the team's decision reasonable and considered?

In this case, the team scores high marks in ethical relevance (the second scale), but low marks for its inclusion of irrelevant information (the third scale). The distinction between commercial and civil law is judged spurious by the judges. They receive average marks in terms of reasonableness (the fourth scale). But the team receives its lowest marks for consistency (the first scale). The judges think that the team cannot justify its belief that the company should support legislation banning such advertisement, given its earlier arguments that allow commerce to act freely and separate from civil society. Each judge fills out a score card and tallies the result. The scores from each judge can range from 0 to 10 points.

The moderator calls on the judges for scoring. With great fanfare each judge raises a card so that the audience and players can see. The team has achieved modest results, scoring three 5s and one 6. Hoping to do better on their next question, they wait until their opponents across the room take their turn.

THE GOALS OF ETHICS BOWL: MORE THAN JEOPARDY, NOT QUITE KANT

The preceding story never actually took place. But it does describe fifteen minutes in the life of a round of Ethics Bowl, created to develop skills needed for ethical reasoning with problems that an adult may face during her professional life. It does so as a supplement to academic learning, providing a venue that can be as substantive as the preparation for it. Designed by Professor Robert Ladenson of the Illinois Institute of Technology, it has been played at IIT annually since 1993 and recently at colleges in Georgia and Texas. In 1997 the first national Ethics Bowl was held during the annual meeting of the Association for Practical and Professional Ethics in Washington, D.C. Fourteen colleges and universities, including Dartmouth College, the University of Texas, and the Air Force Academy, sent teams. The University of Montana's team prevailed.

Unlike most quiz games in which knowledge of a specific kind is rewarded, Ethics Bowl emphasizes the process of ethical reasoning. It rewards and reinforces those activities that contribute to developing considered judgments. It is structured around twenty-five dilemmas from which the questions during the game are drawn. Dilemmas in a wide range of fields are presented: sexual harassment in the Army, personal animosity in the workplace, observing drug use on the job. Each dilemma provides all the details needed to answer the question; no further research is needed.

Here are two Ethics Bowl examples, both written by Ladenson:

1.   You are recently appointed director of a private art museum with one of the finest art collections in the United States. The museum has been losing money steadily for several years, and now has a deficit of over five hundred thousand dollars. The museum's endowment is large enough to cover the current deficit, but it cannot do so indefinitely. A group of active museum board members present to you a plan that calls for selling twenty-five million dollars of valuable French Impressionist artworks in order to create a new endowment fund for "collections care." This new fund will cover costs associated with conservation, preservation, upgraded security, and new acquisitions. The code of ethics of the American Association of Museums forbids selling artwork for purposes other than acquiring more art. The American Association of Museums code of ethics is not legally enforceable.

Should you adopt or reject the plan? In either case, give your reasons.

2.   You are the editor of a scientific journal in the field of abnormal psychology. A paper recently submitted for publication reports on studies conducted at UCLA a decade ago involving schizophrenic patients as subjects.

The results are interesting and valuable from a scientific standpoint. You happen to know, however, that after the research was completed, the National Institutes of Health (NIH) found that the subjects had not been adequately informed of the risks associated with the research. The study, which involved giving medication to fifty patients, and then withdrawing it, resulted in twenty-three patients suffering severe relapses. One of those patients committed suicide, and another tried killing his mother. The investigation concluded that the UCLA scientists conducting the study failed to inform the subjects of the danger of serious relapse. The paper submitted to your journal makes no mention of the findings of the NIH investigation.

What should you, as journal editor, do in this situation and why?

Like most of the questions, these dilemmas are taken from often highly publicized, real events.

The goal of the game is not to trick players but rather to quickly familiarize them with the situations. To this end, the dilemmas are distributed to team members well in advance of the competition. Since judges will query them about their responses, pat answers will not be enough. Players need to consider the other side and offer considered reasons for the positions they take. In the end, it is this ability to understand both sides of the argument and provide reasons for answers that is crucial to game.

THE PREPARATION PROCESS

The intellectual soul of Ethics Bowl is John Stuart Mill's *On Liberty*. In chapter 2, Mill argues that there are four reasons why speech ought not to be restricted to what the majority believes is true:

1. The majority may be mistaken; the other side might hold the true position.

2. Even if the minority is mistaken, its position may be partially true, containing what Mill calls the "moral remainder."

3. Even if the majority opinion is wholly true, it will be reduced to prejudice if its adherents are not forced to "vigorously contest" it.

4. Even if the majority opinion is wholly true, its meaning will be lost. Thus, it will be unable to have a positive effect on the character of the believer, unless the reasons for its belief are fully known to the believer.

Mill had in mind a wide range of thought, from political and religious to the simple social conventions of nineteenth-century England, that to his mind fostered a numbing and dangerous level of social conformity. His approach is one that emphasizes humility—the other side might be right—as well as reason in argument.

Ethics Bowl emphasizes the spirit of Mill through consensus and good reasoning for ethical decision. A team is required to reach a consensus on the answer to any particular question. Although the requirement of consensus is antithetical to Mill, for the purposes of the game, coming to consensus forces team members to consider the strengths of a position they do not hold and to articulate strongly that position—to step in the shoes of those with whom they disagree.

More important, Ethics Bowl emphasizes the reasoning process Mill held so central to deliberation. Good answers are those that are reasoned in a manner that cannot easily be confused with dogma. To achieve this, the four criteria already discussed are given to judges to use in scoring: logical consistency, inclusion of relevant considerations, exclusion of irrelevant considerations, and reasonableness. These criteria are crucial to ethical reasoning in the spirit of Mill and as such take a prominent place in Ethics Bowl.

Logically Consistent.   Students should be introduced to the importance of consistency in making an argument. Good ethical arguments (and most kinds of arguments) do not assert some claim as both true and false simultaneously. Contradictions may make interesting brainteasers, but they do not help make arguments.

Relevant Criteria.   Students should be encouraged to include as much that is relevant to their case and not to leave things out. Even if a student knows that something is relevant but problematic to a particular position, better she should introduce the criteria and explain why it is problematic than to ignore it completely.

Irrelevant Criteria.   At the same time, students should learn what not to include in their arguments. Appeals to emotion, in particular, are often less helpful and more likely to be irrelevant than appeals to reason. Extraneous information should be avoided.

Reasonable Argument.   A student whose argument is logically consistent, includes all the relevant information, and excludes irrelevant considerations, nonetheless must also be reasonable to be a good ethical argument. For example, if I told you that Jane is an inherently better person than John is, and this consideration alone gives her rights to his property, I would be making a logically consistent argument using arguably relevant criteria and not using irrelevant criteria, but still, I would not be persuasive. The reason is that my argument is not reasonable. Note that being reasonable does not mean that everyone has to agree with me. For example, most

Republicans believe that Democrats are *reasonable* even though there are still broad areas of disagreement among them.

VENUES FOR PREPARATION: A CLASS OR A CLUB?

The preparation process allows Ethics Bowl to rise above the level of college "bull sessions" into something more useful and productive. The most substantial preparation can be done using Mill and perhaps one other text on ethical or general reasoning along with the twenty-five questions of Ethics Bowl. The course would be designed primarily for non–philosophy students as an introduction to ethical reasoning. The outline of a twelve-week class might look like this:

Weeks 1 through 4: An introduction to philosophical reasoning
   Reading: Richard Feldman *Reason and Argument** (or other introductory texts on reasoning)
   Assignments: Feldman's book, as do others, includes exercises ideal for this preparation

Weeks 5 and 6: The importance of ethical reasoning
   Reading: J. S. Mill *On Liberty*
   Assignment: short paper or in-class test on the book

Weeks 7 through 12: Twenty-five dilemmas
   Reading: four to five of the Ethics Bowl dilemmas each week
   Assignments: weekly two-page papers providing an argument for one of that week's dilemmas

---

*Richard Feldman, *Reason and Argument* (Englewood Cliffs, N.J.: Prentice Hall, 1993).

During weeks 7 through 12 teachers may want to divide the class into teams of three to five students, and should do so if the class will compete by teams at the end of the semester. Alternatively, the entire class may want to discuss the dilemmas, having only one team of four compete in a university, regional, or national competition. In any case, this last part should include a review of the four already-mentioned criteria of Ethics Bowl listed above.

Preparation for each question is done in groups of three to five not only because of the structure of Ethics Bowl, but because of the role that consensus plays in the preparation. In Ethics Bowl, a single spokesperson must speak on behalf of the team and articulate a unified position. This is admittedly artificial and possibly problematic—ethical commitments and understandings ought not to change in order to reach a consensus, but in this case students may have to adapt.

The twelve-week model is an ideal and thorough way of introducing substance and structure into the preparation. But teams also can be prepared in as little as a week. In these less formal schemes, the four criteria and twenty-five questions are read and discussed. Whether in the classroom or the coffee shop, or somewhere in between, all preparation should include substantial supervision by a faculty sponsor. Indeed, without such supervision and guidance, the freewheeling discussion probably would look like a college bull session.

It is during the preparation stage that the real learning happens. As students grapple to explain their answers to dilemmas and arrive at consensus with their teammates, they will find themselves doing the work of good ethical discourse: providing good reasons for the positions they take and grappling with those who believe otherwise. In cases where immediate consensus exists among the students, it is the advisor's role to play devil's advocate and have the team respond. Similarly, the advisor should ask the team to construct a strong position opposed to that which the team supports.

ALL DRESSED UP AND SOMEWHERE TO GO: PLANNING YOUR OWN ETHICS BOWL

Like the preparation for Ethics Bowl, the actual game can take on a range of forms from an intricate round-robin event involving scores of judges and teams to one in which two teams compete before a panel of two or three judges. No matter what the venue, there are some basic procedures common to all.

Materials Needed

All materials can be created as needed by educators. Alternatively, a packet containing all the materials needed for this game (including a set of twenty-five questions and moderator's answers, rules, and score cards) can be obtained for a nominal fee from the Center for the Study of Professional Ethics at IIT.

- *One "Question Booklet" of twenty-five ethical dilemmas for each team, moderator and judge.* Good situations will be topical (though not necessarily factual), include the necessary information for making a decision, and be five to ten sentences long. Questions should be clearly stated. The scenarios in this volume could easily be developed into one-paragraph questions.
- *One set of twenty-five "Moderator's Answers" for each moderator.* Moderator's answers are used to model a good response, but not meant to be the only "right"

answer. Since they are meant to model good responses, they should not be unduly provocative. However, they may respond in a reasonable but unpopular way.

- *Score sheets for judges.* Score sheets should be divided into the four criteria of judgment: consistency, ethical relevance, ethical irrelevance, and reasonableness. Judges should give scores of 1–5 for each criteria, add the scores, and divide by two. This allows for a range of final scores, 2–10.
- *One set of "Point Cards" (2–10) for each judge.* These are not necessary, but add excitement. When the moderator asks for the judge's scores, they can raise a card with the number of their score, as in Olympic competition.
- *"Rules of the Game" for all participants.* A copy of the rules given in this appendix.
- *Separate tables for each team and the judge's panel.*
- *Podium for the moderator.*
- *Optional prizes (T-shirts, books, et cetera).*

Number of People Needed

- Each team should include three to five students.
- Each team may have a coach, *though this is not necessary for the game.*
- For every two teams there should be three to five judges and one moderator.

Judges should be drawn from a range of fields and should not all be ethics professionals. Judges for past competitions have included lawyers, engineers, graduate students, and an owner of a local grocery store, in addition to college professors.

Schedule Summary

Ethics Bowl lasts approximately two hours (based on a two-round competition). This schedule is designed for four or more teams in a large public setting. Organizers are encouraged to augment the arrangement as best suits their needs.

Schedule Outline

0:00–0:10	Welcome and Explanation
0:10–0:15	Team Division
0:15–1:00	Round One
1:00–1:05	Team Division
1:05–1:50	Round Two
1:50–1:55	Reconvening
1:55–2:10	Wrap-up

Detailed Schedule

0:00–0:10   Welcome and Explanation

After welcoming all guests and participants, the organizer should

- read aloud the "Rules of the Game";
- acknowledge and thank judges, moderators, and team players; and,
- if there are more than three teams, dismiss each pair of teams, three to five judges, and one moderator into a separate room.

0:10–0:15   Team Division

A pair of teams, panels of judges, and one moderator should go to a separate room to begin the game.

0:15–1:00   Round One

Each team should alternately answer two questions as specified in the rules. If time allows, a third or fourth question may be asked.

1:00–1:05   Team Division

Teams should switch rooms so that each is ready to play a second round against a different team. (This requires four or more teams. Smaller groups with only two teams may proceed through more questions.)

1:05–1:50   Round Two

Each team should alternately answer two questions as specified in the rules. If time allows, a third or fourth question may be asked.

1:50–1:55   Reconvening

All team members, coaches, moderators, and judges should reconvene in one location for an announcement of the winner of the competition. If time and interest allow, the top two teams may be given a final question to determine the winner.

1:55–2:10   Wrap-Up

Organizer should wrap up the event by tying the process of the game to both broader issues in ethics as well as everyday ethical reasoning. Prizes, if any, should be given to each team.

Rules of the Game

1. Moderator will decide by coin toss which team goes first.

2.   First team ("Team A") is asked one of the twenty-five questions from the Question Booklet.

3.   Team A has sixty seconds to confer and recall their answer and reasoning.

4.   One member of Team A should then state the answer. This person is then the spokesperson for that question and should be the only one to speak publicly.

5.   Each judge then may ask a brief follow-up question. Questions may require the team to clarify its answer or to discuss briefly an issue which the team's answer raises. If time allows, judges may ask additional questions.

6.   The team may confer for up to thirty seconds on each question asked, after which the spokesperson should respond.

7.   After the judges' questions have been raised and answered, the moderator should state the "Moderator's Answer."

8.   Team A must now choose one of three options:

accept the moderator's answer
accept the moderator's answer with qualifications
challenge the moderator's answer

If either of the last two options are chosen, Team A may have thirty seconds to confer. After conferring, the spokesperson should explain the qualifications with which the team accepts the moderator's answer, or should state the team's reasons for its challenge.

All participants, including team members and judges, should view the moderator's answer as one, but not the only, reasonable answer to the question. The moderator's answer and the team's response to it may be seen as a device to model the element of discussion and dialogue that characterizes reasoning about ethical questions.

9. Each judge must now evaluate the team's treatment of the question on a scale of 1–5. He/she should consider the team's answer, response to the panel's questions, and, when relevant, the explanation of its challenge or qualified acceptance. The evaluation should be based on the following four criteria:

- *Consistency:* Is the team's positions stated and defended in a way that is logically consistent? Is it expressed with enough clarity and precision that the judge can say she or he reasonably understands it?
- *Omissions of Ethical Importance:* Has the team omitted any ethically important considerations?
- *Avoidance of Irrelevance:* Did the team base its position on ethically irrelevant considerations?
- *Reasonableness:* Has the team evaluated the considerations it identified as ethically relevant in a careful and reasonable manner?

Each judge should add scores for the four criteria and divide by two to arrive at a score for the question.

10. At the request of the moderator, judges may in unison hold up their score cards for Team A's answer (in the spirit of Olympic competition).

11. The moderator should add the scores, announce the total, and record it.

12. These steps should then be repeated for the second team with a different question.

CONCLUSION: ENJOYING THE RESULTS

Ethics Bowl was designed to bring a mix of learning and entertainment to what is often perceived as a dull and irrelevant subject. The spectacle of a game of ethics can enliven this subject and help students begin to address the importance of ethical decision making. Past participants, whether players or judges, have praised the "consciousness raising" parts of this process. Depending on the level of preparation, Ethics Bowl also can provide the beginning of ethics education, especially for those students who might otherwise be uninterested and thus less likely to seek out such classes themselves. It provides a modest contribution and innovative supplement to the study of applied ethics.

Index

299